# Practical Statistics

# Practical Statistics

S. S. Cohen
*Senior Lecturer in Applicable Mathematics*
*Polytechnic of Central London*

Edward Arnold
A division of Hodder & Stoughton
LONDON BALTIMORE MELBOURNE AUCKLAND

First published in Great Britain 1988

*British Library Cataloguing in Publication Data*

Cohen, S. S.
Practical statistics
1. Statistical mathematics
I. Title
519.5

ISBN 0-7131-3648-0



Typeset in 10/11 pt Times by J. W. Arrowsmith Ltd.
Printed and bound in Great Britain for Edward Arnold, the educational, academic and medical publishing division of Hodder and Stoughton Limited, 41 Bedford Square, London WC1B 3DQ by J. W. Arrowsmith Ltd, Bristol.

# Preface

It cannot be claimed that there is a shortage of books on statistics. The need is for a text which explains by simple worked examples the techniques which have become indispensable in so many branches of scientific inquiry. Worked examples are used extensively in the teaching of statistics and so a book written in this way should prove a worthwhile addition to the existing literature.

The first three chapters comprise an introduction to the methodology of the subject and form an elementary course for schools or first-year undergraduates. Extensions of this course to intermediate level could include a selection of topics from subsequent chapters like regression, analysis of variance or experimental design. At this intermediate level many statistics books omit work on multivariate data. This seems an unnecessary shortcoming in an age when computers are available and can handle the accompanying arithmetic with ease. This book reflects the trend for degree courses to include multivariate statistics at an earlier stage than hitherto as a result of this development. There is also a chapter devoted to the computational aspects of statistical calculations and the generation of tables of critical values.

A lot of people who need to use statistics find the subject difficult to understand. There is a vast range of procedures for collecting and analysing data and it is not always clear which to select for a specific application. A man with just one watch knows what the time is, a man with two watches is never quite sure. It is hoped that the worked example approach will help to overcome this problem and also enable a researcher to choose which particular questions dealt with in an example are relevant to his or her study. Those who browse through the book as background reading for, say, a management course can likewise delve as deeply as they wish into each of the topics covered.

I want to thank my colleagues at the Polytechnic of Central London for their helpful advice and discussion. All the data values and situations described in the examples are completely fictitious. In this respect I thank my wife Ruth and my children Philip, Elouise, Aaron and Jonathan for listening to my ideas with compassion and for reacting constructively with the upward or downward movement of a thumb.

SSC 1987

# Contents

# 1 Statistical methodology

It is a capital mistake to theorize before one has data—Sherlock Holmes in *Scandal in Bohemia* by Sir Arthur Conan Doyle

Statistics is the art of collecting and analysing sample data in order to gain information about the population from which it was drawn. Pharmacists, for example, have to make decisions about the usefulness of new drugs to the public at large. They base those decisions on data collected from samples of people in what are known as clinical trials. Similar decision making is involved when analysing marketing research surveys, opinion polls, quality control data and other scientific investigations. The complete process by which this is achieved is called statistical methodology and it consists of three distinct stages.

Firstly it is necessary to understand and define the problem in hand, making clear which aspects of it can be quantified and which cannot. For this purpose the objectives of the investigation have to be framed in such a way that statistical techniques can be of help. It is necessary to recognise how and to what extent numerical considerations are relevant. Secondly some **data**, that is a set of measurements or observations, must be obtained which has a direct bearing on the problem as it has been defined. The data may have already been collected for a similar investigation, or the researcher may have to design an experiment, produce a questionnaire or plan some other type of exercise in order to acquire it.

The third stage of statistical methodology is the analysis of the data. This may be as straightforward as drawing a bar chart and calculating some percentages or it may require sophisticated mathematical procedures.

Although the three stages occur consecutively in time they are conceptually interrelated. The data should be collected in a form suitable for the subsequent analysis, which in turn should be appropriate to the objectives defined at the problem identification stage.

We now examine the phases individually and then work through a case study to illustrate the complete process. This also serves as a guide on how to use this book as it refers to ideas and techniques developed in later chapters.

## Defining the problem

### The population

The first task in planning a statistical investigation is to define the population of interest. A **population** is a totality of entities about which we hope to draw conclusions. Each entity can be a person, an animal, an object or an intangible like a road accident. A particular study may relate to one or more than one population. For example we might want to compare the average size of African big cats with that of South American big cats. The two sets of animals would then constitute the two populations of interest.

## Levels of measurement

Having identified the entities to be studied we must decide which of their characteristics are relevant to the investigation. The resulting analysis will describe these in some sort of average sense. A property of a member of a population which varies from one individual to another is called a **random variable**, a **variate** or a **character**. There are several kinds of random variables corresponding to the various levels of measurement which can be achieved in different situations.

The lowest level of measurement is simply to allocate an item to a category depending on the nature of a particular attribute it possesses. For instance, eye colour enables us to classify people and make statements like '23% of the sample had blue eyes'. Observations of such variables are called **nominal, attribute** or **category data** and the number of occurrences, the **frequency**, can be counted for each category.

The next level of measurement has categories which are ranked in some sort of logical order. Attitudes and opinions are canvassed in this way using scales like '1 = agreement', '2 = indifference' and '3 = disagreement' to measure a person's reaction to a given statement. Classifications of this kind give rise to **ordinal data**. Although the categories are in a definite order with 1 and 3 on either side of 2, there is no meaningful 'distance' between them. It does not make sense to say that 'agreement' is 2 units from 'disagreement' just because 3 minus 1 is equal to 2. The scale could have been '1 = agreement', '2 = tendency to agreement', '3 = indifference', '4 = tendency to disagreement' and '5 = disagreement'. The 'distance' now between agreement and disagreement is equal to 4 instead of 2.

If a quantity has the property that subtractions performed on readings do have a physical significance, then a higher level of measurement is being attained. Temperatures are an example and we can make statements like 'it is 4 degrees hotter today than it was yesterday'. Data of this type is called **interval data** because distances or intervals along the scale of measurement are meaningful.

Now a temperature of 20°C does not imply that the weather is twice as hot as when the temperature is 10°C. In contrast, ratios of values for quantities like length, mass and time do have physical interpretations. For instance we can say 'this cat is twice as heavy as that cat'. **Ratio data** like this represents the highest possible level of measurement, because multiplication and division of data values as well as subtraction and addition have meaning. In practice it is rarely necessary to distinguish between interval and ratio data and we shall refer to them jointly as **metric data**. In fact even ordinal data is often treated as if it were metric, possibly after some transformation has been performed on it. Each chapter in this book contains examples relating to nominal and metric data. For situations where there are established methods for dealing with ordinal data they are considered as well.

The classification of a random variable as being nominal, ordinal or metric indicates the level of measurement it achieves. Variables are also characterized by the continuity of the scale on which they are measured. The time taken to boil a kettle of water, say, measured in seconds, might be any number from about 30 to 200. The particular values 62.76, 35.0 and 114.2963 are all possible outcomes for this observation. Such variables are **continuous** and can take any value within a certain range although this can be infinite in extent. In contrast, when there are gaps in the scale between allowable values the variable is **discrete**. The number of caterpillars on a leaf is discrete as it must be a whole number and we cannot observe an outcome like 3.8.

Whenever a continuous variable is measured it generates a corresponding discrete variable. Although length is a continuous variable, a ruler has discrete millimetre markings and so measurements between, say, 23 mm and 24 mm cannot occur. Any measuring instrument, no matter how accurate it claims to be, rounds off the quantity it is measuring at some point. In some situations this phenomenon may need to be borne in mind.

Nominal, ordinal and metric data can be expressed in terms of digits and other characters like decimal points and minus signs. In computing jargon such data is said to be **digital**. A pattern of signals, be they electrical, pneumatic, optical or the like, constitutes **analogue data**. An audio cassette can contain analogue data relating to a piece of music while a photograph is an analogue record of a light pattern. The handling and processing of data in this form is an important branch of computing but for statistical purposes like quality control it is usually converted into digital form first. Pieces of electronics hardware called **analogue to digital converters** and **digital to analogue converters** are used so that a digital computer can read analogue signals and generate control signals in an appropriate analogue form. This is how computers control robots on a production line. As the processing itself is performed on digital data it is usual in statistics to consider this type of data only.

### Factors and treatments

There are often factors present which we know or suspect will cause items within a population to behave differently from each other. These features of the situation may influence the random variables being measured and hence need to be taken into account when planning the investigation.

A **factor** is an effect we can control by setting its **level** before any variables are measured. For example in testing a new drug we would give different doses to different patients, or the same patients on different occasions, before measuring the responses. The various amounts of the drug administered are levels of the factor 'dose'. Notice that dose is not a random variable as its value is completely within our control as part of the design of the experiment.

There can be any number of factors in an investigation or indeed none at all. Each factor can be operative at two or more levels and we define a **treatment** to be a combination of specific levels of each factor present. In the example quoted above it may be that the sex of the patient is considered to be a factor affecting response to the drug. The levels of this factor are 'male' and 'female' and are within our control as we can select specified numbers of men and women for the experiment. There are thus two factors, dose and sex, and combinations like '5 millilitres to a female patient' and '3 millilitres to a male patient' represent possible treatments. Each treatment gives rise to a separate **sample**, this being any set of entities, cases, subjects, items or experimental units chosen from a population. The term is also used to refer to the data values measured on the subjects as well as to the subjects themselves. We assume in this book that all the items in the same sample receive identical treatment in the course of the investigation. If there are separate treatments, then the samples from each one must not be pooled in any way until the analysis phase. By identifying the number of treatments, that is samples, and the number of variables a researcher should be able to locate the chapter of this book appropriate to his or her problem.

## Collecting data

The population or populations of interest are the entities we wish to study. A list of the relevant variables and factors determines the kind of description which will emerge and the next stage of the exercise is to acquire some data on which to base that description. Some researchers find it useful to start with dummy tables of results or dummy diagrams and design the data collection to enable them to fill in the missing figures. On the other hand many studies are purely exploratory and it is neither possible not desirable to anticipate the nature of the final conclusions in this way.

### The law of large numbers

Sometimes it is feasible to examine every population item individually in a **census** or alternatively take a sample which is very large in proportion to the population as a whole. A detailed description of all the items can then be derived without the need for statistical techniques. In most cases, however, the population is too big to be dealt with in this manner and we must content ourselves with a sample which is a relatively small subset of it.

If there is no variability at all between population items then a sample of just one member is sufficient for a complete description of the entire population to be formulated. In this case each identical clone gives perfect information about all the others. Usually of course the population is both large and variable so that any measurement made on only one subject depends on the particular item chosen and is not representative. The impression we get of the population is highly sensitive to the particular item selected.

To see how a sample of more than one item helps us to cope with population variability consider measuring the lengths of 20 cats. Any variability in the sample will, to a certain extent, be 'cancelled out' in the calculation of the average length. Although individual cats can be long or short, the average length of 20 of them will be neither extremely large nor extremely small. This averaging effect becomes more pronounced as the sample size increases but nevertheless requires the sample to be representative of the population as a whole. In essence the **Law of Large Numbers** states that the larger the size of a representative homogeneous sample the better its average estimates the corresponding average of the population. This is the rationale of sampling and the 'safety in numbers' it gives us.

### Replication

The implication of the Law of Large Numbers for statistical methodology is that it is desirable to have as big a sample as possible for each treatment. Each measurement on a separate experimental unit is called a **replication** for the relevant combination of factor levels. Some investigations do not have replication for financial or other reasons and each treatment is applied to just one item. There can also be **repeated measurements** made on the same subject although the analysis of these is different from that of true replication involving separate experimental units.

It is emphasised that the items in a sample must be **homogeneous** with respect to the characteristics we are studying. The extent of this homogeneity determines the accuracy of the resulting description of the population from which the sample was drawn, its **parent population**. A lack of homogeneity will arise, for instance, if a factor within the population is ignored. Suppose that the population of interest is 'all motor vehicles' and we are trying to estimate the average vehicle length. If the sampling process ignores the factor 'car or lorry' and averages all the data together, then the result will be quite meaningless. Clearly separate averages should be quoted for the two types of vehicles. Notice that just because we are able to describe a population easily linguistically, like 'divorced men', this does not in itself imply that there is sufficient homogeneity within the population for any generalisations to be made. The fundamental assumption in a statistical analysis is that members of a sample are identical with each other except for that variability we are prepared to write off as being due to random unexplained variation.

### Sampling methodology

There are various methods of selecting a sample of items from a population. The most common ones used for distributing questionnaires are discussed in Chapter 9 together with the determination of an appropriate sample size. Most statistical techniques are based on the assumption that the sample is **random** and every single member of the parent population has an equal chance of being included in the sample. Lifting spaghetti from a plate with a fork is not random sampling as the longer strands have a higher probability of being selected than the shorter ones. This type of **bias** applies to capturing animals where the older or weaker specimens have a higher probability of being caught. Individual data measurements must be **statistically independent** of each other and ideally should not interact at all.

### Primary and secondary data

A data collection exercise can be costly in both time and money. It is sometimes possible to utilise data which is either from a related study or forms part of a more general investigation like a population census. Data collected specifically for the analysis in hand is called **primary data** while that originally collected for some other purpose is **secondary data**.

Secondary data can be unsatisfactory in several ways. A civil servant reviewing road accidents may define a 'casualty' as someone requiring hospital treatment as an in-patient. An insurance company may call a 'casualty' someone who makes a claim on an insurance policy. The two definitions are different and data obtained by one researcher might be completely useless to the other. Secondary data is also often out of date. Primary data is collected at the time of the investigation and for situations where there is a trend or element of seasonality or topicality, as in an opinion poll, this can be important. Furthermore the sample size for one investigation may be inadequate for another more detailed study. Finally, published results are sometimes biassed to satisfy the needs of the researcher and hence the value of that data in a different study is suspect.

Although on balance primary data is preferable to secondary data, any opportunity to reconcile our results with others should be seized upon. Secondary data, particularly when it relates to a larger sample than the one we have access to, can be useful for this validation purpose.

## Analysing data

By this stage of the investigation we have made observations or otherwise acquired data values on one or more samples, each sample corresponding to a different treatment, which is a combination of factor levels. We now want to draw conclusions about the population or populations of interest and this constitutes the analysis phase of the exercise. In many practical applications of statistics there is too much data and it requires a certain amount of confidence to identify and analyse just that part which is relevant.

Methods of analysis range from the calculation of summary statistics like averages and the drawing of diagrams like bar charts to more sophisticated techniques like analysis of variance and regression. Because the information content of the whole operation is being compressed into a diagram or a simple statement, the result can be like a woman's bikini—what it reveals is interesting but what it covers up is vital. The popular image of statistics has suffered because people like advertisers, politicians and economists have abused its power in order to misrepresent data. Unfortunately there are many ways of telling the truth. For example a sea captain once wrote in the ship's log book 'today the mate was drunk', a perfectly true statement. The following day it was the mate's turn to write in the log and he was naturally upset to read the captain's entry. He wrote the equally true but completely misleading statement 'today the captain was sober'!

## Deterministic models

In a branch of science like chemistry precise statements can be made about the behaviour of individual population items. Chemists predict molecular events from a knowledge of the cause-and-effect relationships between the properties of the substances involved. This description forms a **deterministic model** of the population of interest. In addition, complicated experiments can be carried out to test, one way or the other, the validity of an idea or theory because there is an underlying determinism or causality which always manifests itself. Measurements taken in these experiments are not prone to large random errors and the extreme homogeneity of the populations permits powerful generalisations to be made.

In many disciplines the deterministic level of description implied by statements like 'all chemical X molecules have 4 valency bonds' is not appropriate. In psychology, for instance, such generalisations are either impossible to make or too superficial to have any meaning. The only statements which are true for every single item in the population of interest may be trite or weak like 'every human-being has a brain'. In these areas of knowledge a different form of modelling has to be undertaken.

## Probability

The mathematical way of describing indeterminism is by the concept of probability. A **trial** is an experiment or situation which can have more than one **outcome**. Tossing a coin is an example of a trial with 'heads' and 'tails' being the outcomes. The **probability** of an outcome or set of outcomes is the fraction of times it occurs when the trial is repeated over and over again. We write ***P* (event)** to denote the value of the probability of the 'event' described in the brackets. Thus *P*(a playing card chosen at random is the four of hearts) is equal to 1/52.

The total of the probabilities of all the outcomes of a trial is 1 as they are fractions of the number of trials performed. For instance the probabilities for 'heads' and 'tails' being tossed for a coin add up to 1. The way in which the total probability is 'spread over' the various outcomes is called a **probability distribution** and for a coin this might be 0.5 for the outcome 'heads' and 0.5 for the outcome 'tails'.

Probabilities for metric random variables are often calculated from a **probability density function**. Imagine throwing a needle onto some lined paper and measuring the acute angle it makes with the lines. The angle measured in degrees is equally likely to be any number between 0 and 90 and this uniform density of the probability distribution can be shown on a graph (Fig. 1.1).

The probability of a range of angle values is found by calculating the area under the density graph over the appropriate range. For example, the probability that the needle makes an angle between 25° and 73° is the area hatched in Fig. 1.2.

Hence the required probability is $(73-25)\times(1/90)$, which is 0.5333. This implies that the angle is between 25° and 73° for about 53% of all trials. Note that there is no need to quote probabilities to a great accuracy in final answers. They should be viewed merely as a guide to the likelihood of the event in question and not as highly accurate precise quantities.

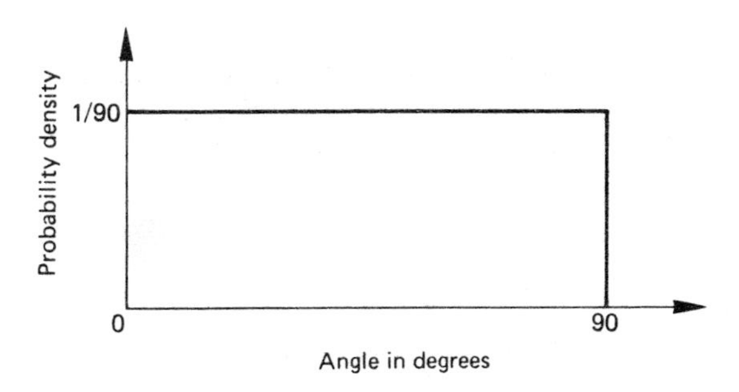


**Fig. 1.1** Probability Density Function for Needle

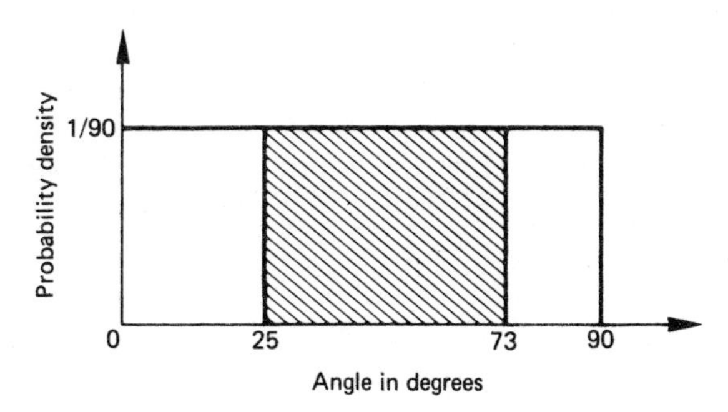


**Fig. 1.2** Area equal to P (angle between 25 and 73)

### Expectation

A probability distribution models the relative frequencies with which various values of a random variable occur throughout the population. The average value of that variable is called its **expectation** and is an important property of the distribution. In the needle example, Fig 1.1 shows that all angles between 0° and 90° have equal probability density. This implies that the average angle will be the mid-point of this range, that is 45°. If we take a sample of angles we would expect its average to be near to this.

When some values of a variable are more likely to occur than others then the expectation will be biassed towards them. Suppose that one quarter of a certain population of plants have one flower on them while the other three quarters have two flowers. Although the population size does not affect the following argument let us suppose further that there are 1000 plants altogether. According to the probabilities,

$$\text{total number of flowers} = \tfrac{1}{4} \times 1000 \times 1 + \tfrac{3}{4} \times 1000 \times 2 \qquad (1.1)$$

and so:

$$\begin{aligned} \text{average number of flowers per plant} &= \frac{\frac{1}{4} \times 1000 \times 1 + \frac{3}{4} \times 1000 \times 2}{1000} \\ &= \tfrac{1}{4} \times 1 + \tfrac{3}{4} \times 2 \\ &= 1.75 \end{aligned} \qquad (1.2)$$

This calculation is unaffected by the population size 1000 and illustrates the general formula:

$$\text{expectation of random variable} = \text{sum of (probability} \times \text{value)} \qquad (1.3)$$

The summation here is over all possible outcomes to the trial and for a continuous random variable it is replaced by an integration. Fortunately, for practical statistics we need understand only the concept of probability density function and expectation. All the ones we use are standard and numerical values can be determined from tables and formulae.

### Stochastic models

Having accepted that a deterministic level of description is inappropriate to a particular investigation, we can try to use probabilities and expectations to build a **stochastic, probabilistic or statistical model**. For a nominal variable this can take the form of a relatively simple statement like 'the probability of a person chosen at random having blue eyes is 0.34'. A set of such probabilities, comprising a distribution over all possible outcomes, can model a population of attributes.

For metric variables the model may be a probability distribution, like the normal distribution in the rice example of Chapter 3, or an equation. For instance, consider the model:

$$\text{height of tree in metres} = \mu_{\text{all trees}} + \text{error} \qquad (1.4)$$

where $\mu$ is the Greek letter 'mu', $\mu_{\text{all trees}}$, a **population parameter**, is the average height of all trees and the term 'error' applies to the specific tree chosen. The equation implies that the height of a tree is made up of a population average which is modified for each individual tree by the presence of a random error. The error can be positive or negative but has an average value of zero for the whole population of trees. It is the error term in the equation which distinguishes it from a deterministic model. We are not attempting to provide a precise prediction for every single item in the population but merely to specify an expectation.

Equation (1.4) has limited meaning because the population of all trees is not sufficiently homogeneous with respect to the variable 'height'. A more detailed description might take into account the factor 'species':

$$\text{height of tree in metres} = \mu_{\text{all trees}} + \alpha_{\text{species}} + \text{error} \qquad (1.5)$$

where $\alpha$ is the Greek letter 'alpha'. In this model each species, like oak or elm, has a separate parameter, $\alpha_{\text{oak}}$ or $\alpha_{\text{elm}}$ to describe the difference between its average height and the average of all the trees.

The second model is preferable to the first because the variability represented by the error term is much smaller. It now stands for variability within a species rather than for the variability amongst the population at large. The error term in a model is a measure of the **residual variability** we are prepared to accept as being unexplained by the model.

Equation (1.5) breaks the population down into more homogeneous subpopulations using the factor 'species'. Of course, within each species the various ages of the trees generates further variability in height. We could try to model this effect:

$$\text{height of tree in metres} = \mu_{\text{all trees}} + \alpha_{\text{species}} + \beta_{\text{species}} \cdot \text{age} + \text{error} \qquad (1.6)$$

where $\beta$ is the Greek letter 'beta'. In this model the parameter $\beta_{\text{oak}}$, for example, is the increase in the average height of oak trees for each year of their age. It multiplies the age of a tree to give an 'age contribution' to the height. As with most organisms, however, trees do not continue to grow at the same rate throughout their lives. It might be better to model some function of their height, like the logarithm, in order to obtain a realistic description.

The particular model assumed in the analysis of a given set of data values depends on the detail and nature of the data collected. This highlights the interrelationship between the analysis phase of an investigation and its initial planning phase. If the objective of the operation is to arrive at a model like (1.6) with two factors, species and age, then clearly heights must be measured for trees of different species at various ages. On the other hand, if a model specifically for willow trees is required, then only heights of willow trees at different ages would have to be measured. Age is then the only factor in the investigation.

Most models are, or can be transformed into, a sequence of constant terms denoting factor levels, like species, and multiples of variables, like the age term in (1.6). They are called **linear models**.

The question we must answer next is 'how does the data collected in an investigation enable us to build a statistical model and test its validity?' Before examining this problem we look at the more general process by which scientific theories absorb the knowledge content of observations.

## Scientific method

Scientists try to formulate theories which not only explain their own particular observations but also provide generalisations about the subject matter in hand. Our present knowledge of astronomy and medicine, for example, is the culmination of thousands of years of data collection. There are many statements, models or 'laws' in these sciences which are the result of successive refinements of earlier theories tempered by a multitude of observations. Each new theory has to explain all the accumulated knowledge to date as well as the observations on which it is based.

The process of **deductive inference** occurs when a specific is derived from a generalisation, a 'minor' from a 'major'. For example if I take the statement 'all robins have red breasts' to be true then I can say 'the next 3 robins to land in my garden should all have red breasts'. The whole of mathematics is dependent on deductive reasoning. The axioms of arithmetic, geometry and algebra are generalisations and individual calculations or proofs of theorems are specifics derived by deduction from those statements. Deduction is merely the exposing to view of results which are implicit within the original axioms. This procedure on its own can add nothing to the power of the axioms themselves as truths about the population of interest. Schoolchildren have been doing arithmetic for hundreds of years but they have not and cannot discover an essentially 'new' rule by the process of deduction alone. Science needs a mechanism whereby the generalisations themselves can be updated in the light of fresh and revealing observations.

Framing generalisations from specifics is called **inductive inference** and involves deriving a 'major' from a 'minor'. An example of this would be to observe that 249 robins all have red breasts and inducing that 'all robins have red breasts'. No claim is made that the generalisation represents the absolute truth. It is merely the current theory which we hold to be 'true' because all the observations made to date are consistent with it and suggest its validity. Such a 'law' is rejected if just one piece of evidence emerges which contradicts it. The appearance of a single robin with a breast which is not red disproves conclusively the truth of the law 'all robins have red breasts'. The cycle of theory, contradictory evidence, new theory, new contradictory evidence and so on is called **scientific method**. All branches of science claim to progress in this way. It is the leap from data to theory which characterises the induction process and this must at some point be incorporated into a statistical analysis.

### Statistical inference

The application of the induction process to the data from a sample, which is the 'minor', to a model of the population, which is the generalisation or 'major', is called **statistical inference**. The values of the population parameters in a model can either be estimated from the data or made the subject of a hypothesis which is to be accepted or rejected.

*Method 1: estimation*
The value of a population parameter in a model can be estimated using a **sample statistic**. For instance the mean of the sample, which is such a statistic, gives an indication of the mean of the entire population. If the population mean was a parameter in our model we could therefore take the sample mean as an estimator of it. Sample statistics are themselves random variables as they depend on the particular sample selected.

The role of probability in statistics can now be summarised. The population has variability which we wish to describe. Probabilities give us a tool to do this with and we include random error terms in our models to represent background 'noise'. That same variability means that different samples yield different sample statistics and hence different estimates of population parameters. Probability distributions help with this problem by enabling us to calculate a range of values within which we believe, with a stated degree of confidence, the population parameter lies. We can also evaluate expectations of sample statistics to decide whether they are true or biassed estimators of corresponding population quantities. The theory behind the estimation process is considered in more detail in the Squirrels Example of Chapter 6.

*Method 2: hypothesis testing*
The second approach to statistical inference is to adopt specific values for population parameters and examine whether the data is consistent with them or not. We accept a hypothesis about a population parameter unless the data constitutes evidence which is strong enough to disprove it. As in scientific method generally, disproof rather than proof plays a dominant role. Again probabilities enter the discussion and allow us to assess the likelihood of an experimental result when compared with an assumed hypothesis. Every sample gives a different picture of the behaviour of the population. In testing public opinion we might, just by chance, find an entire sample of people who all agree with government policy even though the national percentage of such people is, say, only 83%. A quality control inspector might select, just by chance, a sample of good components from an essentially bad batch of items or, conversely, a bad sample from a good batch. Either way he is misled by the information he obtains, but in our decision making we can set the value of the probability of the risk we are prepared to take.

We have now considered the three phases of a statistical investigation, the planning stage, the data collection stage and the analysis stage. The chapter ends with an example to illustrate them in a practical setting.

### The piglets example

As usual, the farmers of Ditchpuddle are spending the evening in the village inn talking about pigs. Soon enough the conversation turns to the weight of six-month-old male piglets who were born in the United Kingdom. The farmers decide to weigh their own piglets in order to explore the subject further.

*Question 1* Identify the elements of a statistical investigation as they apply to this situation.

*Solution* The population of interest consists of all six-month-old male piglets born in the United Kingdom. The single random variable being measured is 'weight' which will yield metric data.

There are numerous factors which might account for the variability in weights throughout the population. Amongst other things, one would suspect that variety of pig, size of litter and age of mother all affect the growth rate and ultimate size of a male piglet. A more detailed study would include several samples corresponding to the different levels of these factors. A picture of how different treatments influenced the weight of the piglets would then emerge.

The data collection strategy the farmers adopt is simply to examine every six-month-old male piglet born in Ditchpuddle. This would constitute a census if the population of interest was 'all Ditchpuddle piglets' but gives a biassed view if the farmers are concerned about the national scene. Piglets born in other parts of the country have no chance of being included in the sample, which is therefore not a random one.

*Question 2* The farmers return to the inn the following night and pool their data. There are a total of 27 six-month-old male piglets in the village and their average weight is 104.7 kg. What can the farmers infer about the national average weight?

*Solution* In pooling their data, the farmers are ignoring yet more factors about the situation, in particular the specific farm on which the piglet was reared. In accepting the data as being a single homogeneous sample, the farmers are proposing the model:

$$\text{weight of six-month-old male piglet in kg} = \mu + \text{error} \quad (1.7)$$

This implies that the expected average of any random sample of piglets is $\mu$, the population mean weight. Taking their particular sample mean as an estimate of this quantity, **the farmers conclude that the mean weight of six-month-old male piglets born in the United Kingdom is about 104.7 kg.**

*Question 3* The farmers notice that their individual measurements are quite different from each other. They calculate that the standard deviation is 3.1 kg. How does this help in the estimation process?

*Solution* The standard deviation is described in Chapter 2 and its use in estimation in Chapter 3. All that concerns us here is that a measure of the spread of the data values in a sample enables us to qualify the degree of confidence we have in an estimate. The details of the calculation are in the Train Journey Example of Chapter 3 and in this case **the farmers can be 95% certain that the national average is between 103.5 kg and 105.9 kg.**

*Question 4* An eminent pig breeder writing in the *Pig Farmers' Clarion* claims that the national average male piglet market weight is 98.4 kg. Does the Ditchpuddle data contradict this claim?

*Solution* As well as being processed to estimate population parameters, data is also used to test hypotheses about the population. The pig breeder's hypothesis that the

mean is 98.4 kg presents us with a dilemma as this value lies outside the interval we have 95% confidence in. The problem can be summarised as follows:
*A The pig breeder is right.* This implies that the farmers' calculations are wrong somewhere along the line or they have been unlucky in collecting a sample with a probability smaller than 5% of occurring by chance.
*B The pig breeder is wrong.*

The philosophy of statistical hypothesis testing is to assume that the sample is not one of an extreme 5% group but is unbiassed and representative. **We therefore conclude that the population mean weight is not 98.4 kg.** This is a shaky conclusion in the present example as the data has so many obvious shortcomings. Unfortunately, statisticians are often in the position of having to work with unsatisfactory data. The topics of estimation and hypothesis testing are covered in more detail in Chapter 3.

## Summary

**1** The first phase of a statistical investigation is the identification of the population or populations of interest together with the variables which will be measured.

**2** Data can be nominal, ordinal or metric depending on the level of measurement which can be achieved. Variables can be discrete or continuous.

**3** In the data collection phase it may be necessary to acknowlege the existence of factors which affect population behaviour. The number of factors studied and the levels of each factor determine how many distinct treatments are possible. Each treatment gives rise to a single measurement observation or a sample of replications. The Law of Large Numbers implies that the larger the sample size for each treatment the better.

**4** Primary data is collected especially for the investigation while secondary data has been used elsewhere but has relevance to the current problem. It should be utilised in checking conclusions wherever possible.

**5** In the analysis phase statistical models are built by means of probabilities and expectations. Statistical inference is a special case of scientific method in which either the values of population parameters are estimated or hypotheses about them are tested.

## Further reading

Most of the books listed in the first four sections of the Bibliography contain a discussion on the methodology behind a statistical investigation. References (5) and (7) are particularly good while books on nonparametric methods like (33) and those written for specific disciplines like (14) examine different levels of measurement in more detail than other texts. Sampling and data collection techniques are dealt with in Chapter 9 of this book. The references on questionnaire design and sample surveys include much useful material in these areas. In fact, just as data collection tends to receive too little attention in most general statistics text books, so probability theory attracts too much. Probabilities are a tool in a statistical analysis like algebra and calculus. It is not necessary to understand a lot of theory to utilise probabilities to describe the randomness of selecting a sample and the degree of confidence to be placed on its message. Of course there is an underlying theory to the probability distributions of statistics and this is covered in books like (21) to (26) and (28). References (20) and (27) of the Theoretical Statistics section present an alternative method of statistical analysis to the one used in this book. **Bayesian statistics**, as it is called, views the collection of data as an information gathering exercise which modifies our knowledge of the relevant probability distributions which describe the situation.

Statistical methods are employed in a multitude of scientific and not-so-scientific disciplines. This range of applications is reflected in the section on 'Statistics for Specific Disciplines' which is by no means completely comprehensive. Incidentally, reference (31) has been written especially for social scientists. For researchers who envisage utilising secondary data as discussed in this chapter, references (49) and (50) provide good starting points in identifying suitable sources.

The ideas of deterministic and stochastic modelling are used in subjects related to statistics like operational research. Reference (55) gives an excellent treatment of probability and the application of the two types of modelling in this kindred branch of applicable mathematics.

## Exercises

**1** Suggest possible populations of interest, variables to be measured and suitable objectives for each of the following researchers: (i) a psychologist, (ii) a civil engineer, (iii) an astronomer, (iv) a sociologist, (v) a balletomane, (vi) a marine biologist and (vii) yourself.
**2** Suggest data collection methods for the above. Would secondary data be of use? Describe how the cycle of theory, data, new theory applies in each case.
**3** Classify the following variables as being nominal, ordinal or metric levels of measurement. State also whether they are discrete or continuous: (i) salary, (ii) blood type, (iii) the number of penguins on an island, (iv) the acid concentration in a sample of urine, (v) the position a horse finishes in a race, (vi) the number of births per thousand people in the population, (vii) an earthquake rating on the Richter scale and (viii) a person's favourite food.
**4** An airline passenger is afraid that the aeroplane he is due to travel in will be blown up by a bomb. He discovers that although many single bombs are found on aircraft, it is extremely rare for two or more bombs to be found on the same flight. He therefore decides to take a bomb with him. Comment.
**5** Captain Nemo swings his spaceship into the flight path of the two enemy craft. He knows full well that one has 3 Thark missiles and a Wombat decoy and the other only one Thark and two Wombats. If he chooses a ship at random and chooses a missile on the ship at random, to destroy, what is the probability that it is a Thark?

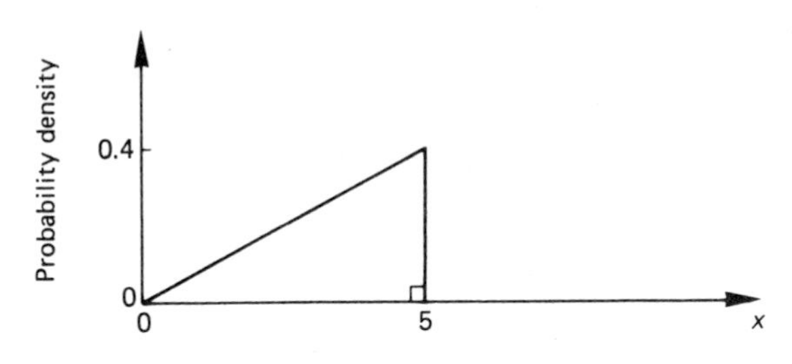


**Fig. 1.3** Probability Density Function

**6** The continuous random variable $x$ has the probability density function shown in Fig. 1.3. Calculate the probability that (i) $x$ is less than 2, (ii) $x$ is between 3.5 and 4.5 and (iii) $5x$ is bigger than 15.

7 Our senses and hence our perceptions seem to operate in just three spatial dimensions. Data, statistical or otherwise, can therefore reach our intellect in this form only. The theory of relativity and some ideas in particle physics suggest that the universe has more than 3 spatial dimensions and there is no logical reason why it should not have an infinite number. Man's perceived universe is thus merely a three-dimensional cross-section of the real thing, a shadowy image. Discuss how this consideration restricts the processes of scientific and statistical method. Can we still hope to build mathematical models which completely describe the universe? (Good answers to this question should be sent to the Nobel Prize Committee mentioning the author's name.)

# 2 Describing data

A statistical analysis is about extracting information from data. This procedure often starts, and often ends as well, with the drawing of diagrams or the calculation of various statistics. Diagrams display the overall characteristics and information content of a sample. They appear as aids to communication in newspapers, books, reports, television programmes and advertisements. **Summary statistics** are numerical descriptors of the sample and can be used either to complement diagrams or on their own. By making explicit its main features a preliminary analysis can also act as a diagnostic for the suitability of further techniques.

## Bar charts, pictograms and pie charts

There are many different types of diagrams and it is important to choose one which is appropriate to the sample in hand. The end product should be self-explanatory to the reader and communicate effectively. It should have a meaningful title and, if applicable, axes which are clearly labelled with the names of the variables they represent and their units of measurement.

### The game reserve example

A game warden sees 46 elephants, 34 antelopes, 13 giraffes and 23 zebras on his reserve. Illustrate these observations by means of suitable diagrams.

*Solution* The data is nominal and although the relative percentages of the different species observed can be calculated there are no obvious numerical measures for the

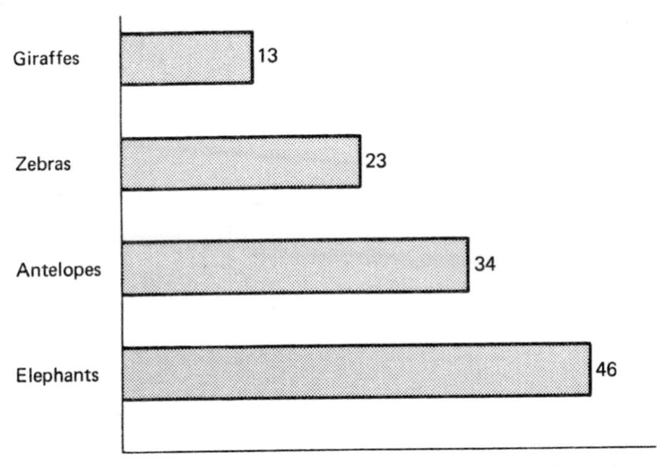


**Fig. 2.1** Population sizes for the Game Reserve Example

sample. **Bar charts** like the one in Fig. 2.1 are popular for showing the relative occurrence of an attribute. The lengths of the bars represent the frequencies with which the various categories were found in the sample. It is important that all the bars have the same width and that their entire lengths are drawn because, given the choice, the human eye/brain system responds to areas rather than lengths. The diagram in Fig. 2.2 is misleading because the areas, which have more visual impact than lengths, are not in the ratio of the frequencies they are supposed to represent. For example the impression is given that there are the same number of giraffes as zebras. Similarly the areas in Fig. 2.3 give an incorrect account of the data as the frequency scale has been shortened with the zero suppressed. Here there appears to be three times as many zebras as giraffes.

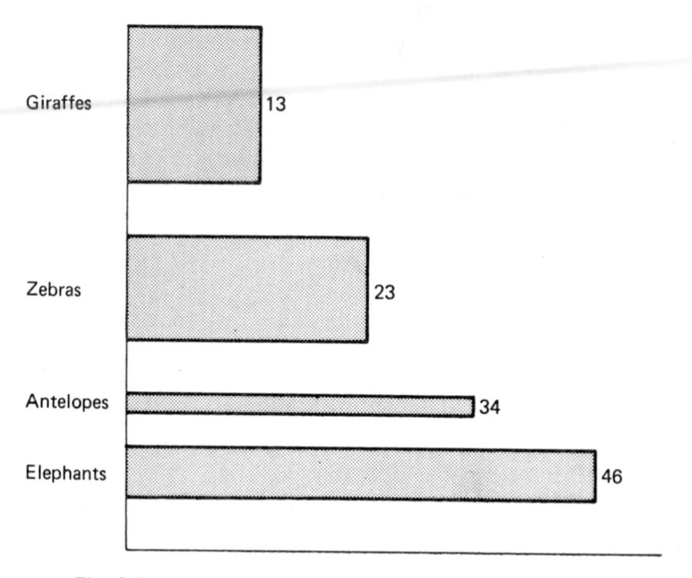


**Fig. 2.2** Misleading Diagram I—different width bars

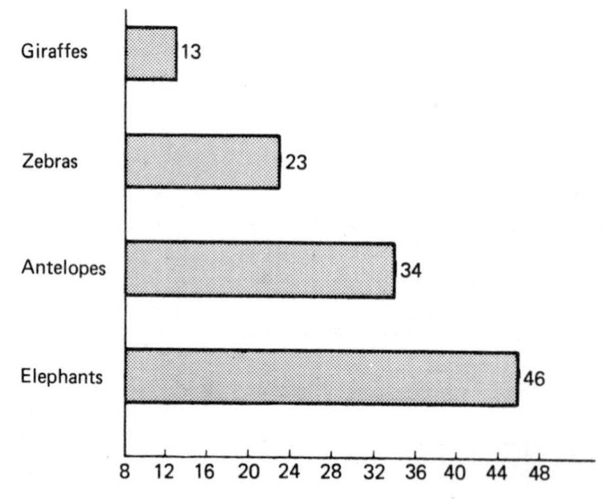


**Fig. 2.3** Misleading Diagram II—suppression of zero

When the diagram uses volumes instead of lengths or areas to create an impression of the data then there is more scope for deception. Even though the lengths of the sides of the cubes in Fig. 2.4 are proportional to the corresponding frequencies, the volumes convey a false meaning. For instance, the number of elephants appears to be much more than twice the number of zebras because no less than 8 'zebra cubes' fit into the 'elephant cube'. A more accurate picture would show the apparent volumes rather than the lengths in the proportion of the frequencies being displayed.

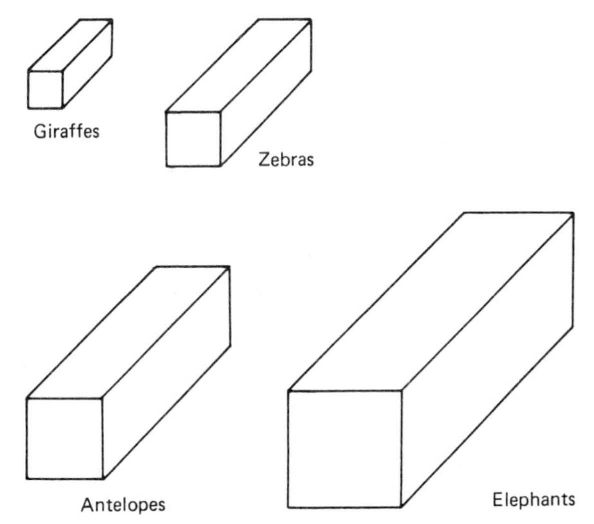


**Fig. 2.4** Misleading Diagram III—different apparent volumes

A **pictogram**, Fig. 2.5, has a number of little pictures to illustrate each frequency. Several pictures are used rather than a single one to overcome the eye/brain perception problem described above resulting from apparent volumes.

Sometimes the frequency for a category of an attribute can be subdivided within that category. Suppose in the present example we knew the sexes of the various animals

**Fig. 2.5** Pictogram of the animals in the Game Reserve

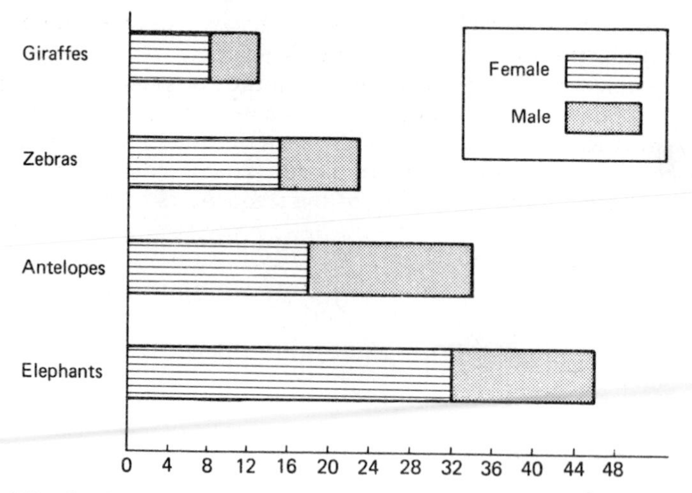


**Fig. 2.6** Stacked Bar Chart of populations for the Game Reserve Example

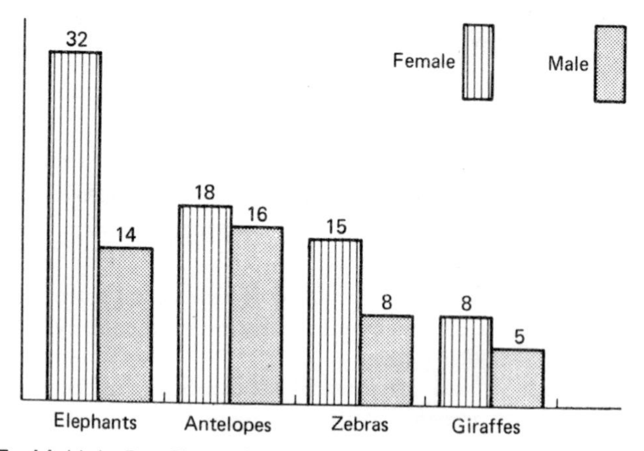


**Fig. 2.7** Multiple Bar Chart of populations for the Game Reserve Example

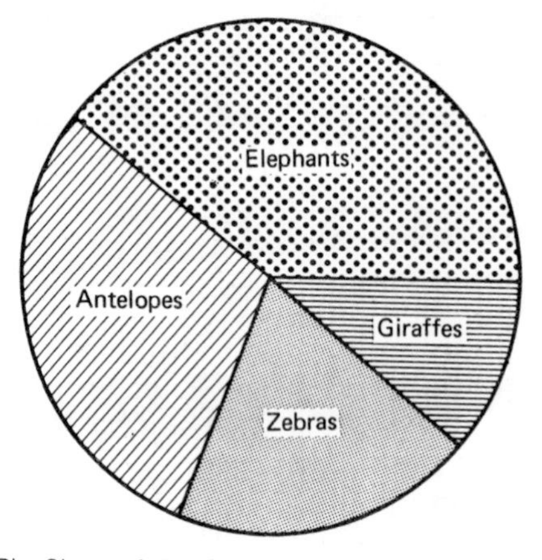


**Fig. 2.8** Pie Chart of the Game Reserve Example populations

observed. Figure 2.6 is a **stacked bar chart** showing this information while Fig. 2.7 is a **multiple bar chart** with the bars drawn side by side.

When the data can be thought of as being a breakdown of some whole entity into component parts then a **pie chart** (Fig. 2.8) can be drawn. This applies to our example if the four species observed are the only ones found on the reserve. The total 'pie' then corresponds to the complete set of all the animals and the slices or sectors represent the sizes of the various subpopulations. In a pie chart the angle of each sector is proportional to the frequency with which that category of data occurs. A pie chart is sometimes drawn in an **exploded** form with its sectors spaced apart from each other or in elevation as an ellipse. Most of the diagrams described in this example are available on computer statistics and spreadsheet packages. They make it very easy to produce multi-coloured bar charts and pie charts with all the necessary calculations and elevations being done automatically.

## Numerical measures for raw data

### The shower example

The times, measured in minutes, taken by 15 people to have a shower were:

3.8, 4.2, 8.1, 7.0, 6.3, 5.4, 4.9, 6.7, 6.3, 5.7, 6.3, 4.5, 5.0, 5.4, 4.4.

We summarise the information content of this metric data sample numerically.

*Solution* There are several ways of describing the overall size of the data values. The most common measure of position, as they are called, is the **mean** or **average**:

$$\text{Mean, } \bar{x} = \frac{\text{sum of data values}}{\text{number of data values}} \tag{2.1}$$

In this example it is:

$$\bar{x} = \frac{3.8+4.2+8.1+\cdots+4.4}{15} = \frac{84}{15} = 5.6 \tag{2.2}$$

In quoting an average we must remember its units. **This sample has an average of 5.6 minutes.**

The average of a sample does not always give a true picture of the overall position of the data. Extraordinarily large or small $x$ values, called **outliers**, will have a disproportionate effect on it. This phenomenon is especially apparent in economic data like salaries, where one or two very high values in a sample will distort the mean. The **median** is a measure which is not unduly affected by any single data value as they all carry the same 'weight' in its calculation. It is defined to be the middle value when all the data is arranged in numerical order. If the sample size is even and there is no middle number, then it is the average of the middle two.

The shower times written in numerical order are: 3.8, 4.2, 4.4, 4.5, 4.9, 5.0, 5.4, 5.4, 5.7, 6.3, 6.3, 6.3, 6.7, 7.0, 8.1 and so the middle one, the 8th, is 5.4. **The median of the sample is 5.4 minutes.** The median is such that half the sample values are smaller than it and half are larger. The mean has this property only if the data is symmetrically distributed about some central value. The median is often applied to data measured on an ordinal scale.

In some situations neither the mean nor the median are appropriate. The manager of a shoe shop may not be interested in the average or median size of the shoes that he sells but in the **mode** or most popular value. **The mode of the shower data is 6.3 minutes** as this value occurs more frequently than any other. If all the data values occur equally often, then the sample has no mode while there can be more than one mode if several of them have the highest frequency. Of the three measures of position defined above only the mode applies to nominal data. For the Game Reserve Example the most common animals to be observed were elephants and so that category is the mode.

We now examine the extent to which the shower data values are spread apart. This is an important property of a sample in an application like quality control. It is no consolation to a manufacturer that the average length of the bolts he makes is on target when individual bolts are too small or too big to be used.

The simplest measure of spread is the **range**. This is equal to the largest data value minus the smallest one. **In the Shower Example it is 8.1 − 3.8 which is 4.3 minutes.** Although it is quick and easy to calculate, the range is highly sensitive to outliers in the sample. It is usually preferable to have a measure which reflects the spread of the whole sample rather than just its most extreme members.

The **variance**, $s^2$, is a measure of spread which is related to the deviations of the data values from their mean. There are three equivalent formulae:

$$s^2 = \frac{\text{sum of }(x^2) - ((\text{sum of }x)^2/n)}{(n-1)} \tag{2.3}$$

$$s^2 = \frac{\text{sum of }(x^2) - n.\bar{x}^2}{(n-1)} \tag{2.4}$$

$$s^2 = \frac{\text{sum of }((x-\bar{x})^2)}{(n-1)} \tag{2.5}$$

where $n$ is the size of the sample.

The numerator of these formulae is called the **corrected sum of squares** as it is the sum of the squared $x$ values after the mean has been subtracted from them. **Centring** the data, as this is called, is carried out by many computer packages as a prelude to further analysis and so (2.5) is often used in them. Formula (2.3) on the other hand requires just one 'pass' over the data and is used in scientific calculators. Both the sum of the $x$ values, often called '$\Sigma x$' on the keyboard, and the sum of the squared values, '$\Sigma x^2$', are evaluated together as the data is entered in.

For the shower data formula (2.3) gives the result:

$$s^2 = \frac{(3.8^2 + 4.2^2 + 8.1^2 + \cdots) - (84^2/15)}{(15-1)}$$

$$= \frac{490.08 - 470.4}{14} = 1.405\,71 \tag{2.6}$$

Hence **the variance of the shower times is 1.41 minutes²**. The units of variance are the squares of the original ones because the variance is a function of the data values squared. It is common practice to take the square root and obtain a measure called the **standard deviation,** $s$. This has the same units as the data. In our example:

$$\text{Standard deviation, } s = \sqrt{\text{variance}} = \sqrt{1.40571} = 1.1856 \tag{2.7}$$

and **the standard deviation of the times is 1.19 minutes.**

Care should be taken over the use of formulae (2.4) and (2.5). Accuracy may be lost when $\bar{x}$, which is the result of a division, is rounded off. This imprecise value may then distort the calculation of $s^2$. We shall use version (2.3) almost exclusively as it incorporates only the totals of the raw data. On the subject of accuracy, notice that although a generous number of decimal places are retained in the calculation itself, answers are quoted to a realistic level of precision. The shower data is given in the question correct to one decimal place. It would be wrong to claim an accuracy for the mean, variance and standard deviation which is out of proportion to this fact. Thus the answers have been rounded off to two decimal places.

Most scientific calculators have special facilities for determining the mean and standard deviation of a sample. There are usually two standard deviation options, one marked $\sigma_{n-1}$ or $s$, which corresponds to our formulae and is the **sample standard deviation**, and another marked $\sigma_n$ or $\sigma$, the **population standard deviation**. The latter applies only when the data relates to the entire population and the formulae for it then have $n$ in the denominator of (2.3)–(2.5) instead of $n-1$. We shall have no use for it as all our work is with samples which are not complete populations.

The reader might be interested in why the denominator of the sample variance formulae is $n-1$ and not $n$. After all, it is the average squared deviation of the data values about their mean. Consider the sample 4, 8, 15. It contains 3 separate pieces of information. However, after the mean has been calculated, a knowledge of any two data values together with the mean enables the missing third one to be reconstructed. In other words, as the mean is 9 the sum of all the data must be 27 and so subtracting any two of them from this yields the third one. Hence calculating the mean of a sample reduces its information content from $n$ to $n-1$. The variance is said to have $n-1$ **degrees of freedom** as it depends on $n-1$ independent values. In the extreme case where as many statistics have been calculated as there are data values, then the number of degrees of freedom is reduced to zero and the information content of the sample has been completely drained.

Sometimes the size of the numbers in a sample gives rise to computational problems. Their sum and the sum of their squares may be large and lead to numerical errors. It is helpful in such cases to subtract a suitable number, an **assumed mean**, from every data value. The mean and variance of the adjusted data are calculated as before and the mean readjusted by adding back the assumed mean afterwards. For instance if the original data values were around 200 we might decide to subtract 200 from every one of them. If the resulting average is 33.6 with variance 7.42, then the mean of the original sample is 233.6 with the same variance 7.42. The range, variance and standard deviation all remain the same for the adjusted sample as its spread is precisely the same as that of the original data.

The next example illustrates the calculation of measures of position and spread for data in the form of a frequency table.

## Frequency tables



### The bananas example

The numbers of bananas on each of 218 bunches were counted and the data converted into a **frequency table**:

| Number of bananas | 5 | 6 | 7 | 8 | 9 |
|---|---|---|---|---|---|
| Number of bunches (frequency) | 90 | 65 | 38 | 17 | 8 |

The information in this table could be represented as a bar chart. We shall concentrate on describing it numerically and determine the mean, median, mode, range, variance and standard deviation of the number of bananas on a bunch.

*Solution* The table shows that there are 90 data values equal to 5, 65 equal to 6, 38 equal to 7 and so on. This repetition must be taken into account when we add them together and add their squares together. The working for this can be tabulated (Table 2.1).

**Table 2.1** Sum of $x$ and sum of $x^2$ for the Bananas Example

| $x$ value | Frequency $f$ | $f \cdot x$ | $f \cdot x^2$ |
|---|---|---|---|
| 5 | 90 | 450 | 2250 |
| 6 | 65 | 390 | 2340 |
| 7 | 38 | 266 | 1862 |
| 8 | 17 | 136 | 1088 |
| 9 | 8 | 72 | 648 |
| Total | 218 | 1314 | 8188 |

The first two columns are copied from the question. The third column lists the contribution to the total of all the $x$ values due to each frequency. For example, as there are 38 data values all equal to 7, their contribution to the total of all the $x$ values is 38 times 7, which is 266. The last column shows the calculation of the sum of the squares of the data. The contribution of the 38 repetitions here is 38 times $7^2$, that is 1862.

Having determined the sum of the $x$ values and the sum of their squares in the bottom row of the table, we can calculate the mean, variance and standard deviation:

$$\bar{x} = \frac{\text{sum of } x}{n} = \frac{1314}{218} = 6.0275$$

$$s^2 = \frac{\text{sum of }(x^2) - ((\text{sum of } x)^2/n)}{(n-1)}$$

$$= \frac{8188 - (1314^2/218)}{(218-1)} = \frac{267.8349}{217} = 1.2343$$

$$s = \sqrt{1.2343} = 1.1110 \tag{2.8}$$

Thus **the mean is 6.03 bananas, the variance is 1.23 bananas$^2$ and the standard deviation is 1.11 bananas.**

The median of a sample is the middle value when they are all arranged in numerical order. In this example there are 218 items, an even number, so we take the average of the two middle values. From the frequency table in the question we see that the first 90 values in the sample are equal to 5, the next 65 are 6 and so on:

$$\underbrace{5, 5, 5, 5, \ldots, 5, 5,}_{\text{90 numbers}} \underbrace{6, 6, 6, 6, \ldots, 6, 6, 6,}_{\text{65 numbers}} \underbrace{7, 7, 7, 7, \ldots, 7, 7,}_{\text{38 numbers}} \underbrace{8, 8, 8, \ldots, 8, 8,}_{\text{17 numbers}} \underbrace{9, 9, 9, \ldots, 9}_{\text{8 numbers}} \tag{2.9}$$

It follows that the middle two numbers, the 109th and the 110th, are both equal to 6. Hence their average is 6 and **the median of the sample is 6 bananas.**

**The mode**, being the most common values, **is 5** as this number has the highest frequency of occurrence. Finally **the range**, which is the largest data value minus the smallest, is $9-5$, that **is 4 bananas.**

## Grouped data, histograms and ogives

When the sample contains many different values a frequency table is of little help. Such data can be classified in order to convey information about its nature.

### The dandelions example

The lengths of the stalks of 205 dandelions were measured and classified to form a **grouped frequency table**:

| Stalk length (mm) | 1.0–4.0 | $4.0^+$–6.0 | $6.0^+$–8.0 | $8.0^+$–10.0 |
|---|---|---|---|---|
| Number of dandelions (frequency) | 9 | 25 | 34 | 40 |

| Stalk length (mm) | $10.0^+$–12.0 | $12.0^+$–14.0 | $14.0^+$–20.0 | $20.0^+$– |
|---|---|---|---|---|
| Number | 42 | 35 | 16 | 4 |

The plus sign in this table means 'greater than' so the class '$8.0^+$–10.0' includes all dandelions whose stalks are *greater than* 8 mm but *less than or equal to* 10 mm. The class labelled '$20.0^+$–' contains all cases for which the length was greater than 20 mm. For computational purposes we shall have to assume an upper limit on the dandelion stalk lengths, say 25 mm. The first and last classes tend to have relatively small frequencies and so the choice is not usually critical.

When classifying data it should be remembered that although detailed knowledge of the figures is being lost, the frequency table ought to convey information about the sample as a whole. The classes used in the grouping process should be chosen to achieve this.

*Question 1* Display the data by means of a suitable diagram.

*Solution* The diagram must illustrate not only the frequencies with which the data values occur within the different classes but also the sizes of the class widths. A **histogram** (Fig. 2.9) is a kind of bar chart where the widths of the bars represent the widths of the classes. As areas have more visual impact than lengths we draw bars whose areas are proportional to the frequencies for each class (Table 2.2).

**Table 2.2** Calculations for the histogram of the Dandelions Example

| Class width (mm) | 3 | 2 | 2 | 2 | 2 | 2 | 6 | 5 |
|---|---|---|---|---|---|---|---|---|
| Frequency | 9 | 25 | 34 | 40 | 42 | 35 | 16 | 4 |
| Frequency/width (height of histogram) | 3.0 | 12.5 | 17.0 | 20.0 | 21.0 | 17.5 | 2.7 | 0.8 |

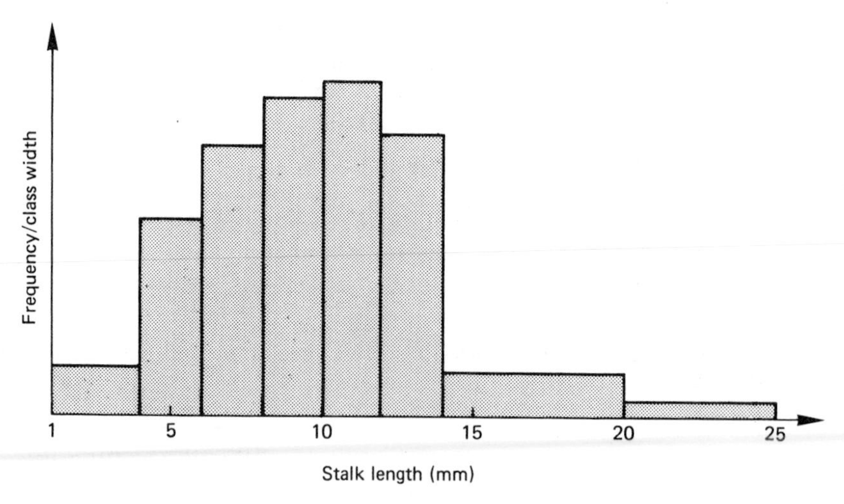


**Fig. 2.9** Histogram of the Dandelion Stalk lengths

Calculating the heights in this way ensures that the area of the resulting rectangle illustrates the frequency. The scale of the vertical axis and its label, 'Frequency/class width' are often omitted, leaving the areas to indicate the relative sizes of the frequencies without conveying their exact values.

*Question 2* Calculate the mean, variance and standard deviation of the data.

*Solution* The sum of the data values and the sum of their squares cannot be evaluated accurately as we do not have access to the original measurements. In order to estimate these quantities, we assume that each dandelion has a stalk length equal to the mid-point of the class to which it belongs. For instance the 25 lengths in the range 4 mm to 6 mm are all assumed to be equal to 5 mm. As in the last example the working can be tabulated (Table 2.3).

**Table 2.3** Working for the Dandelion Example mean and variance

| Class | Frequency ($f$) | Mid-point ($x$) | $f \cdot x$ | $f \cdot x^2$ |
|---|---|---|---|---|
| 1.0–4.0 | 9 | 2.5 | 22.5 | 56.25 |
| $4.0^+$–6.0 | 25 | 5.0 | 125.0 | 625.00 |
| $6.0^+$–8.0 | 34 | 7.0 | 238.0 | 1 666.00 |
| $8.0^+$–10.0 | 40 | 9.0 | 360.0 | 3 240.00 |
| $10.0^+$–12.0 | 42 | 11.0 | 462.0 | 5 082.00 |
| $12.0^+$–14.0 | 35 | 13.0 | 455.0 | 5 915.00 |
| $14.0^+$–20.0 | 16 | 17.0 | 272.0 | 4 624.00 |
| $20.0^+$–25.0 | 4 | 22.5 | 90.0 | 2 025.00 |
| Total | 205 | — | 2 024.5 | 23 233.25 |

As in Table 2.1, the first two columns are copied from the given data but in this case an upper limit has been supplied for the last class. Formulae (2.1), (2.3) and (2.7)

give:

$$\bar{x}=\frac{2024.5}{205}=9.8756 \qquad s^2=\frac{23\,233.25-((2024.5)^2/205)}{(205-1)}$$

$$=\frac{3240.078}{204}=15.8827$$

$$s=\sqrt{15.8827}=3.9853 \tag{2.10}$$

Hence **the mean is 9.88 mm, the variance is 15.88 mm$^2$ and the standard deviation is 3.99 mm.**

There is a **correction due to Sheppard** which can be applied to compensate for the use of the mid-points instead of the original data values. It consists of subtracting $((\text{class width})^2/12)$ from the variance. It can be used only if all the classes have the same width.

*Question 3* Estimate the number of data values in the original ungrouped sample which were between 11.5 mm and 15 mm.

*Solution* The **cumulative frequency** of $x$ is the number of data values which are less than or equal to $x$. It is an especially helpful concept when describing grouped data. In our sample, although the raw data has been classified, the exact cumulative frequencies of certain stalk lengths can be deduced with accuracy (Table 2.4).

**Table 2.4** Cumulative frequencies for the dandelion sample

| Stalk length (mm) ($x$) | 1 | 4 | 6 | 8 | 10 | 12 | 14 | 20 | 25 |
|---|---|---|---|---|---|---|---|---|---|
| Cumulative frequency (number of values less than or equal to $x$) | 0 | 9 | 34 | 68 | 108 | 150 | 185 | 201 | 205 |

The entry corresponding to a stalk length of 12 mm, for instance, is the sum of the frequencies 9, 25, 34, 40 and 42 in the grouped frequency table. We know that all these stalks had lengths less than or equal to 12 mm because of the classes they have been put into. Similar calculations are performed for the right-hand end-point of all the classes to obtain Table 2.4. Note that the cumulative frequency of 1 is zero as there are no data values less than 1 mm.

For a continuous random variable, cumulative frequency increases smoothly as $x$ increases. We expect the number of dandelions with stalk lengths less than or equal to, say, 11.8 mm to be very slightly bigger than the number whose lengths are less than or equal to 11.7 mm. The extra 0.1 mm in length will add those between 11.7 mm and 11.8 mm to the cumulative frequency and this should be a relatively small quantity. Thus a graph of cumulative frequency against $x$ should be a smoothly varying curve. It is shown in Fig. 2.10 and is called a **cumulative frequency curve** or **ogive**. The graph provides estimates of the cumulative frequency for $x$ values in between the ones plotted from Table 2.4. It is therefore possible to take readings from the curve to reconstruct properties of the original unclassified data. In particular we can derive the estimate asked for in the question.

On an accurate version of Fig. 2.10 the cumulative frequency of 15 is about 190 and that of 11.5 is around 138. Now stalks bigger than 11.5 mm but less than or equal to

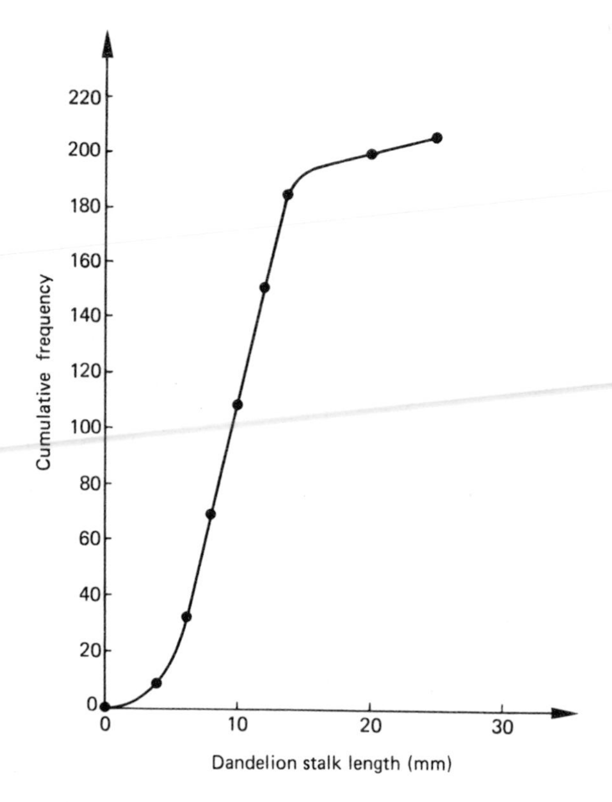


**Fig. 2.10** Ogive of the Dandelion Stalk lengths

15 mm will contribute to the cumulative frequency of 15 but not to that of 11.5. Hence the number of stalks between 11.5 mm and 15 mm is estimated by subtracting 138 from 190. **We believe there were about 52 dandelions with stalks lengths between 11.5 mm and 15 mm.**

*Question 4* What can be said about the mode of the data?

*Solution* Having no knowledge of the raw values, we estimated the mean, variance and standard deviation of the data by assuming they are distributed uniformly within each class. For the same reason we cannot estimate the mode except to remark that it corresponds to the $x$ value for which the ogive has the steepest slope. This is because the biggest increase in the cumulative frequency will occur as $x$ passes through the mode. It is difficult to estimate from the graph and sometimes the **modal class** is quoted for grouped data. **This is the class with the highest frequency and for the dandelion sample is '10 mm to 12 mm'.**

*Question 5* Use the ogive to describe the sample numerically.

*Solution* There is a collection of summary statistics associated with the ogive and the concept of cumulative frequency. It relates to the distribution of the data values along their range (Fig. 2.11).

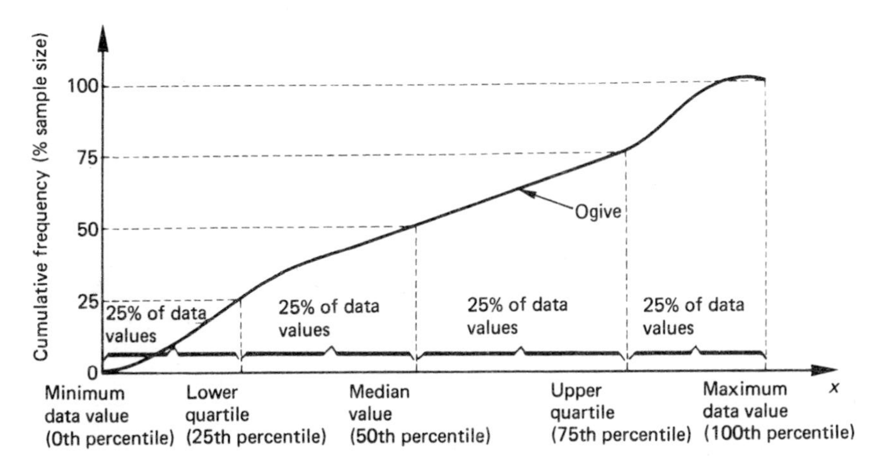


**Fig. 2.11** Numerical measures associated with Cumulative Frequency

The quantities along the $x$ axis split the range into 4 sections. Their values can be read from Fig. 2.10 in accordance with Table 2.5.

**Table 2.5** Ogive measures of position for the Dandelion Example

| Quantity | Cumulative percentage | Cumulative frequency (sample size is 205) | $x$ value which has this cumulative frequency from Fig. 2.10 |
|---|---|---|---|
| Minimum | 0 | 0.0×205=0.00 | 0.0 |
| Lower quartile | 25 | 0.25×205=51.25 | 7.0 |
| Median | 50 | 0.50×205=102.5 | 9.5 |
| Upper quartile | 75 | 0.75×205=153.75 | 10.4 |
| Maximum | 100 | 1.00×205=205.00 | 25.0 |

The measures listed in Table 2.5 are special cases of **percentiles**. The 64th percentile, for example, has the property that 64% of the data values are less than or equal to it. The 10th, 20th, 30th, etc. percentiles are called **deciles**. Economists make statements like '10% of the working population spend less than £5 per week on entertainment'. This is another way of saying that the first decile, the 10th percentile, of amounts spent on entertainment is £5.

The **lower and upper quartiles** in Table 2.5 provide a measure of the spread of the data. Half the values in the sample are between these two numbers and their distance apart is the **interquartile range**. To give an idea of the variability either side of the median it can be halved to produce the **semi-interquartile range**. For samples which are symmetric about the median, half the data values are between (median − semi-interquartile range) and (median + semi-interquartile range).

A description of the sample based on the ogive is therefore that **25% of the data is less than the lower quartile (7 mm), 25% is above the upper quartile (10.4 mm), the median is 9.5 mm and the semi-interquartile range is 1.7 mm.**

## Time series, frequency polygons and indices

A sequence of measurements corresponding to different instants or periods of time is called a **time series**. It should be noted that a time series is a single sample with two variables being measured, the quantity of interest and time. Here is an example.

### The record player example

A shop makes the following sales of record players:

| Year | 1980 | 1981 | 1982 | 1983 | 1984 | 1985 |
|---|---|---|---|---|---|---|
| Sales of record players | 342 | 285 | 381 | 357 | 362 | 344 |

We shall display the data with a suitable diagram and discuss its indexation.

*Solution* The histogram in Fig. 2.12 has the years as classes and the sales figures as frequencies. The year-by-year changes in sales can perhaps be better illustrated by a **frequency polygon** (Fig. 2.13).

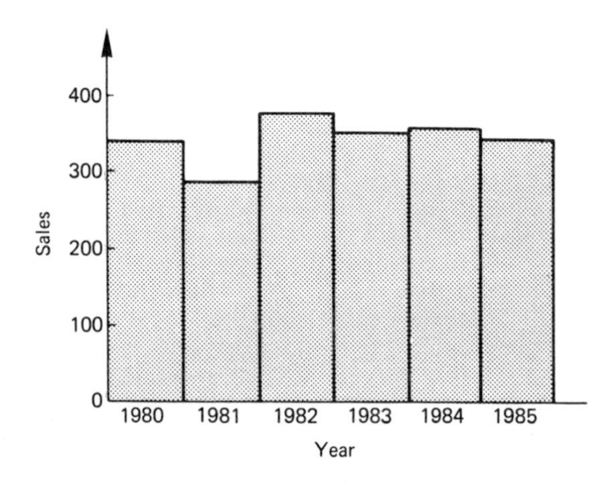


**Fig. 2.12** Histogram of Record Player Sales

This consists of straight lines joining the mid-points of the tops of the rectangles in a histogram. Straight lines are used because a curve gives the impression that all the points in between those plotted have some sort of meaning. This is not true for such graphs as a vertex of the polygon represents a yearly total and so relates to a period of time rather than to a specific time instant. A frequency polygon is really a graph of the histogram heights against the classes they describe.

It is often felt necessary to focus attention on the movement of the figures in a time series rather than on their absolute size. An **index** is established for the series by expressing each data value as a percentage of the figure for a **base** year. In our example if the base year is 1980 then the indexed level of sales for 1983 is (357/342) expressed as a percentage, that is 104. A frequency polygon of these indices would look exactly like Fig. 2.13 but with a different scale on the vertical axis.

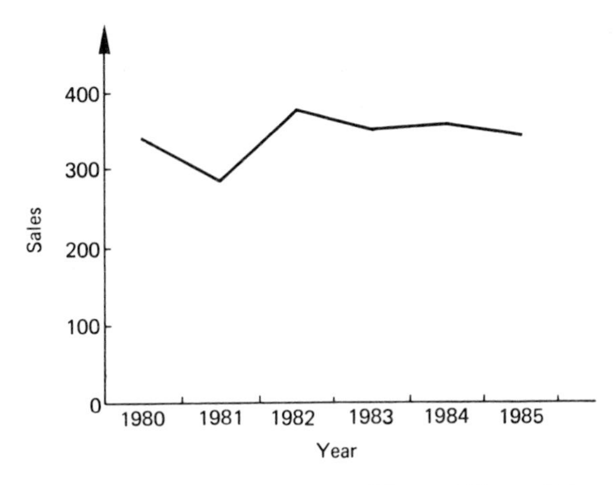


**Fig. 2.13** Frequency Polygram of Record Player Sales

## Graphs and scatter diagrams

Scientific data is often represented by a **graph**. Different ways of examining a sample with more than one variable in this manner are covered in the next example.

### The gulls example

The numbers of male and female gulls found on a desert island on the same day of the month over an eight month period were:

| Month | Mar | Apr | May | Jun | Jul | Aug | Sep | Oct |
|---|---|---|---|---|---|---|---|---|
| Number of male gulls | 84 | 92 | 88 | 86 | 91 | 114 | 119 | 107 |
| Number of female gulls | 66 | 81 | 110 | 125 | 115 | 124 | 113 | 117 |

*Question 1* Draw a diagram to show the behaviour of these populations.

*Solution* Although they are both time series, the data above has a different nature from that of the Record Player Example. The changes from one time period to another of an aggregated sales figure can be quite abrupt. A bar chart of frequency polygon was used in preference to a smooth curve so that the impression of continuous variation was not conveyed. In the present example, population size is a quantity which varies gradually with the passage of time. It is not the accumulation of an effect over a period of time. When two quantities change continuously and smoothly with respect to each other, a graph can be plotted of their values. The gull population data gives rise to Fig. 2.14. Occasionally a logarithmic scale is used so that lengths represent multiplicative or percentage change. Other transformations of the data peculiar to its context may help in exposing the specific behaviour patterns the researcher is seeking.

*Question 2* Draw a diagram to show the relationship between the male and female populations of the gulls.

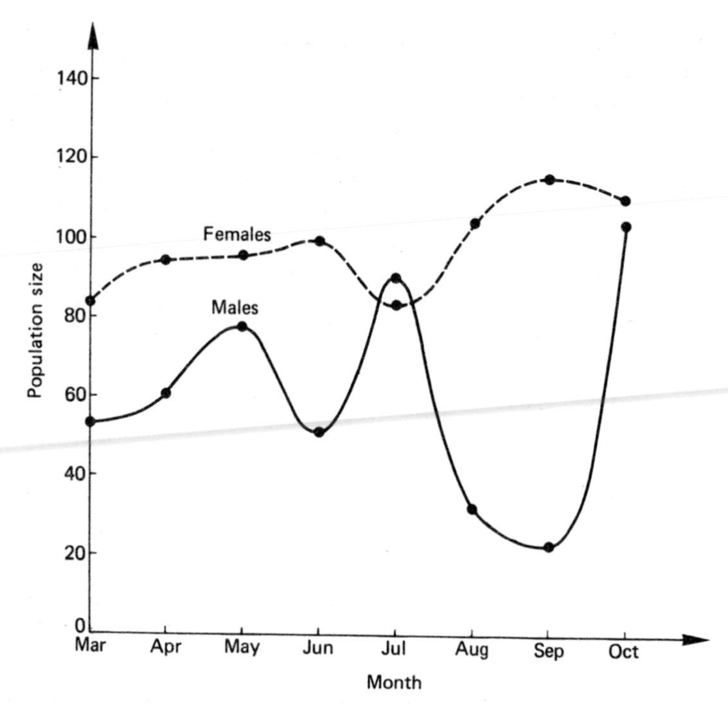


**Fig. 2.14** Population sizes of male and female gulls

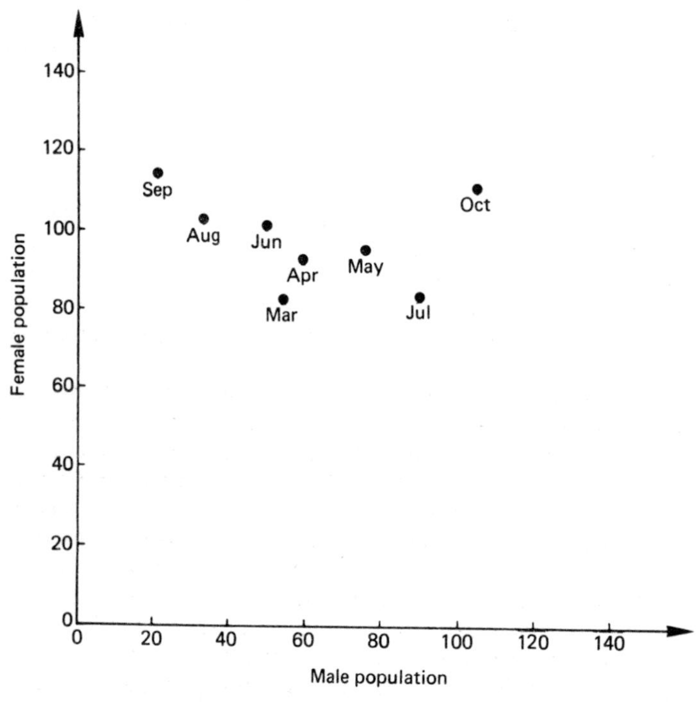


**Fig. 2.15** Scatter Diagram for the Gulls Example

*Solution* Just as Fig. 2.14 displays the relationship between each of the population sizes and time, so a graph can be used to illustrate the relationship of the population sizes with each other. The result (Fig. 2.15) is called a **scatter diagram** as it shows how the points representing each item in the sample are scattered in accordance with the two variables being measured. The data values are used as coordinates in this diagram and any pattern the points may have gives us information about the relationship between the variables being measured. In this example there is no clear evidence of a correlation between the two populations. A more comprehensive treatment of the interpretation of scatter diagrams is contained in Chapter 4. Suffice it to say here that the search for a causal connection between two variables often begins and ends with the drawing of a scatter diagram like Fig. 2.15 which is merely an amorphous cloud of points.

## Summary

**1** Bar charts, pictograms and pie charts give descriptions of the frequencies of the different categories for nominal variables. It is important when drawing them to be aware of the visual impression being created as they can be misleading.

**2** The mean, median and mode of a sample are measures of its position. The range, variance and standard deviation measure the spread of the data values about their mean.

**3** Data values in a sample can be summarised as a frequency table or a grouped frequency table. An ogive can be drawn for the cumulative frequencies of grouped data which allows properties of the original sample to be reconstructed. There are various measures of position and spread associated with an ogive.

**4** Frequency polygons and indices can help in analysing a time series. Graphs are generally used for scientific data. A scatter diagram is a plot of one variable against another showing how, if at all, they are related.

## Further reading

All the books in the first two sections of the Bibliography have material on the drawing of diagrams and the calculation of summary statistics. Sometimes it is called exploratory data analysis and more diagrams and computing techniques are given in (56). Most spreadsheet and statistics packages for computers allow the user to generate diagrams and calculate measures of position and spread. They also facilitate the sorting and classifying of raw data.

The analysis of time series, which were introduced in this chapter, and forecasting are specialised branches of statistics and operational research. References (60) and (65) are texts dedicated to this subject while (55) has a chapter on it. Care should be taken with series of monetary amounts. Those relating to separate times should be discounted to allow for differences in real value before being compared with each other.

Multivariate data is often represented by cluster diagrams or trees and these are dealt with in Chapter 5. Cross-tabulations and contingency tables for two variables are used in Chapter 4, as are scatter diagrams in correlation analysis.

The importance of the mean and variance in statistics stems from the fact that their values characterise a normal distribution. Classical sampling theory, which assumes normally distributed populations, is developed in the next chapter and this importance is made apparent.

## Exercises

**1** A sample of 285 holidaymakers were asked about the principal mode of transport they used to travel to their holiday destination. The results were:

| Method of transport | Aeroplane | Boat | Car | Coach | Train |
|---|---|---|---|---|---|
| Number of people | 84 | 15 | 135 | 92 | 66 |

Represent the data (i) by a pie chart and (ii) by a pictogram.

**2** The reigns of Keith I, II, III, IV, Bert IX and Queen Enid were for 13, 8, 22, 16, 11 and 25 years respectively. (i) Represent the data by a suitable diagram and (ii) calculate the mean, median, range, variance and standard deviation of the sample.

**3** The sound intensity levels measured in decibels at 81 construction sites were 68, 63, 59, 77, 60, 57, 63, 62, 64, 73, 63, 60, 70, 71, 65, 65, 68, 67, 67, 62, 56, 61, 69, 64, 58, 73, 68, 66, 65, 64, 68, 67, 68, 67, 69, 70, 69, 61, 65, 62, 68, 62, 69, 64, 82, 66, 65, 69, 68, 70, 62, 64, 64, 65, 68, 69, 71, 62, 68, 63, 63, 69, 69, 67, 64, 64, 63, 66, 68, 69, 63, 63, 62, 61, 67, 61, 64, 65, 65, 68, 70. (i) Compile a frequency table for the classes '50–60', '$60^+$–65', '$65^+$–70', '$70^+$–75' and '$75^+$–85'. (ii) Draw a histogram of the data. (iii) Calculate the exact mean and variance of the sample from the raw data values. (iv) Estimate the mean and variance of the data from the grouped frequency table by assuming each value is equal to the mid-point of the class to which it belongs. (v) Compare your answers to parts (iii) and (iv) and comment on the validity of the classification scheme.

**4** The duration times of 726 telephone conversations were:

| Length of call (seconds) | 0–200 | $200^+$–400 | $400^+$–500 | $500^+$–600 | $600^+$– |
|---|---|---|---|---|---|
| Number of calls (frequency) | 40 | 208 | 253 | 193 | 32 |

(i) Draw an ogive for the data. (ii) Estimate the median, the upper and lower quartiles and the semi-interquartile range from the graph. (iii) Estimate the probability that a telephone conversation chosen at random lasts less than 450 seconds. (iv) Estimate the probability that a telephone conversation chosen at random lasts between 450 seconds and 550 seconds.

**5** A company achieves the following levels of sales in the years 1978 to 1985:

| Year | 1978 | 1979 | 1980 | 1981 | 1982 | 1983 | 1984 | 1985 |
|---|---|---|---|---|---|---|---|---|
| Sales (£) | 42 500 | 43 800 | 46 400 | 45 600 | 46 300 | 46 100 | 45 700 | 46 200 |

These figures have been discounted to allow for the change in the value of money over the relevant time period and are therefore directly comparable with each other. (i) Plot a frequency polygon of the series. (ii) Calculate the mean and standard deviation of the sample by using an assumed mean of £45 000 and working in units of £1000. (iii) Taking 1978 as the base year with an index of 100, calculate indices for the sales figures in subsequent years. (iv) The advertising expenditure of the company over the same period was:

| Year | 1978 | 1979 | 1980 | 1981 | 1982 | 1983 | 1984 | 1985 |
|---|---|---|---|---|---|---|---|---|
| Advertising (£) | 5100 | 4200 | 4900 | 4500 | 5200 | 4700 | 4800 | 5000 |

Plot a scatter diagram to show the amount of advertising against the level of sales. Comment on the presence or absence of a relationship between the two variables.

# 3 One sample with one variable

The techniques considered in this chapter apply when just one observation is made on each item in a single random sample. It is important that every experimental unit is treated in exactly the same way so that the data values obtained are directly comparable with each other and the sample is homogeneous. There must be no factors which subdivide the population in some fashion and so give rise to essentially separate samples, each one corresponding to a different factor level.

The chapter is in four sections. The first deals with sampling from populations which the normal probability distribution can be assumed to describe. In view of the Central Limit Theorem discussed in the second section, this is not as restrictive a condition as it may appear to be. In the next section, distributions are introduced for $t$ and $\chi^2$, pronounced 'ki squared' to rhyme with 'pie squared', $\chi$ being a letter of the Greek alphabet. They form the basis of what is known as classical sampling theory and are used in estimation and hypothesis testing.

The third section is about the description of data by the uniform, binomial, Poisson and normal probability models. They are tested for validity with the $\chi^2$ statistic. The last section of the chapter is concerned with some non-parametric methods. These enable conclusions to be drawn about populations for which the use of averages or other metric parameters is not appropriate or, for one reason or another, the normality assumption of classical sampling theory is invalid.

## The normal distribution

The use of histograms to depict the frequencies of grouped data was demonstrated in Chapter 2. Many metric variables generate histograms having a distinctive bell-like

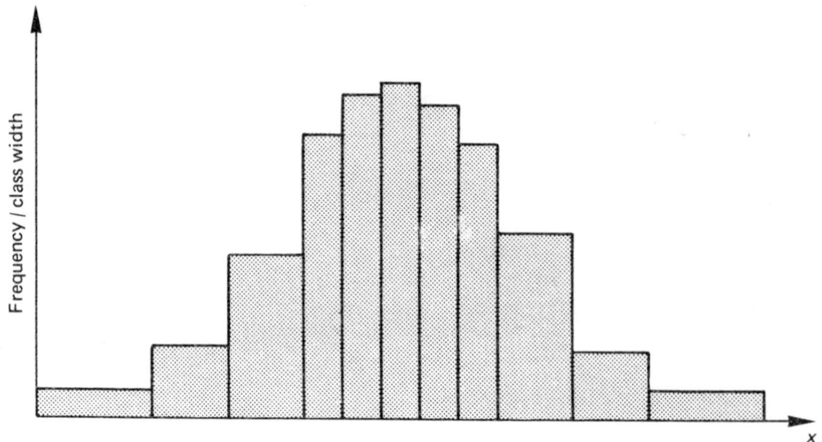


**Fig. 3.1** Histogram of a sample drawn from a normal population

shape. Heights or weights, for example, tend to be concentrated around a central average, with large deviations from that average being relatively uncommon. The resulting histogram is therefore peaked in the middle with two symmetric 'tails' on either side, as in Fig. 3.1. Many years ago the designation 'normal' was given to the shape of such histograms because data which did not conform to it was considered to be 'abnormal' in some sense.

A normal probability distribution models this type of situation and is characterised by two parameters, the population mean, $\mu$, and the population standard deviation, $\sigma$. The symbols $\mu$ and $\sigma$ are the Greek letters 'mu' and 'sigma' respectively. The mean is the value of the random variable where the peak in the curve occurs while the standard deviation measures the spread or flatness of the graph. In practice we never know the values of these quantities exactly but must estimate them from the data in a sample. Probability distributions were introduced in Chapter 1 and the following example shows how this particular distribution is applied.

**The rice example**

In a food processing factory, rice is packed in bags bearing the label '1.5 kg'. As part of the quality control system, each bag is weighed as it leaves the production line and its weight is recorded. The production manager draws a histogram of the weights of many thousands of bags and sees that it resembles Fig. 3.1. He therefore accepts the normal curve as an adequate description of the distribution of the weights. As the sample is very large its mean, 1.56 kg, and its standard deviation, 0.4 kg, are taken to be good estimates of the corresponding population values, $\mu$ and $\sigma$. The suitability of the normal distribution to a particular situation can in fact be tested and this is demonstrated in the Concrete Example of Chapter 10.

*Question 1* Use the normal distribution to calculate the probability that a bag of rice chosen at random from the production line weighs more than 1.8 kg.

*Solution* As frequency is represented by area on a histogram, it corresponds to the area under the appropriate normal curve. A sketch of the curve is shown in Fig. 3.2

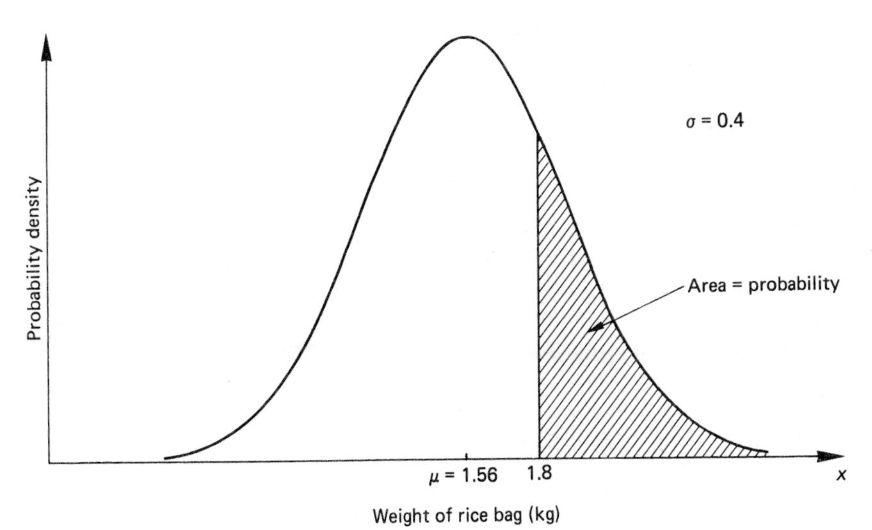


**Fig. 3.2** Normal curve for Question 1 of the Rice Example

and the shaded area denotes those bags of rice in the population weighing more than 1.8 kg. To determine the probability of choosing one of these at random we need the size of this area relative to the area under the whole curve. This is equivalent to finding an area under the **standard normal curve**, which is a scaled version of all other normal curves. The scaling relationship is

$$z = \frac{x - \mu}{\sigma} \tag{3.1}$$

and the new curve has a total area equal to one square unit. Formula (3.1) is called the **z score** of the original $x$ value.

For the present example we obtain the diagram in Fig. 3.3, which is a sketch of what we have done so far in the calculation. Areas under this curve are tabulated in statistics texts or books of statistical tables and those on page 183 of this book indicate that the area we want is 0.27425. Rounding this off to two significant figures, **the probability that a bag of rice chosen at random weighs more than 1.8 kg is 27%.**

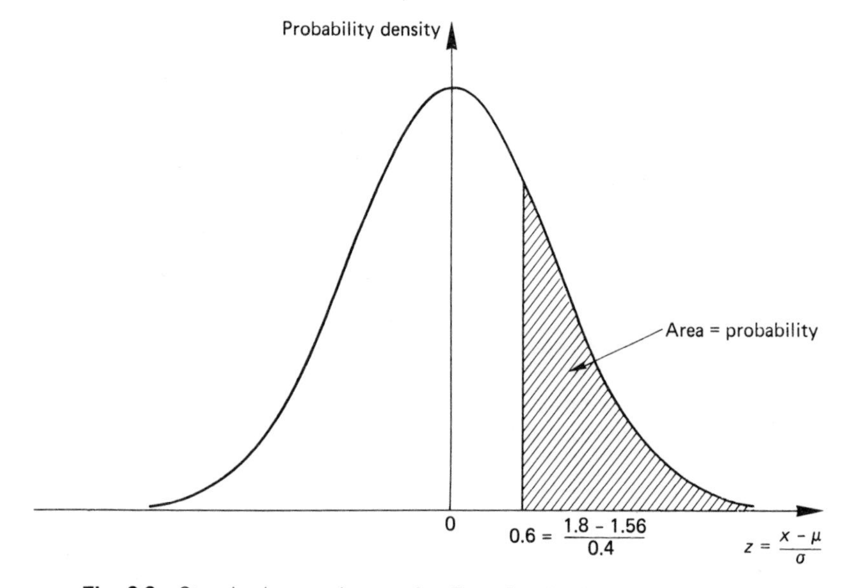


**Fig. 3.3** Standard normal curve for Question 1 of the Rice Example

Readers with a knowledge of calculus may be wondering why we do not integrate the equation of the standard normal curve in order to evaluate areas beneath it. Unfortunately, the function, which is quoted on page 182, cannot be integrated in closed analytical form. The computer program on page 185 is a numerical method of finding the area by summing a power series. The next few questions illustrate the use of tables obtained in this way.

*Question 2* What is the probability that a bag chosen at random weighs less than 1.62 kg?

*Solution* The '$x$ diagram' is shown in Fig. 3.4, and this scales to the '$z$ diagram' (Fig. 3.5). Referring again to the tables on page 183 we find that the area to the right of

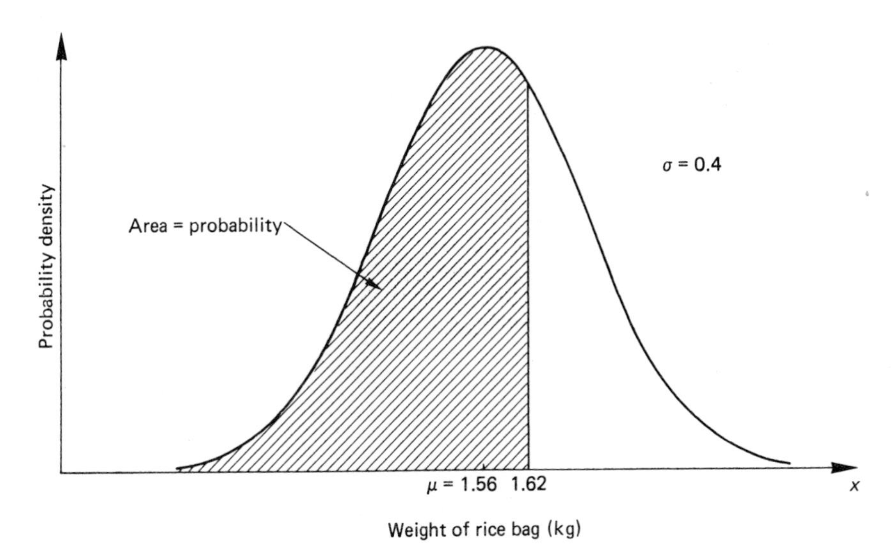


**Fig. 3.4** Normal curve for Question 2 of the Rice Example

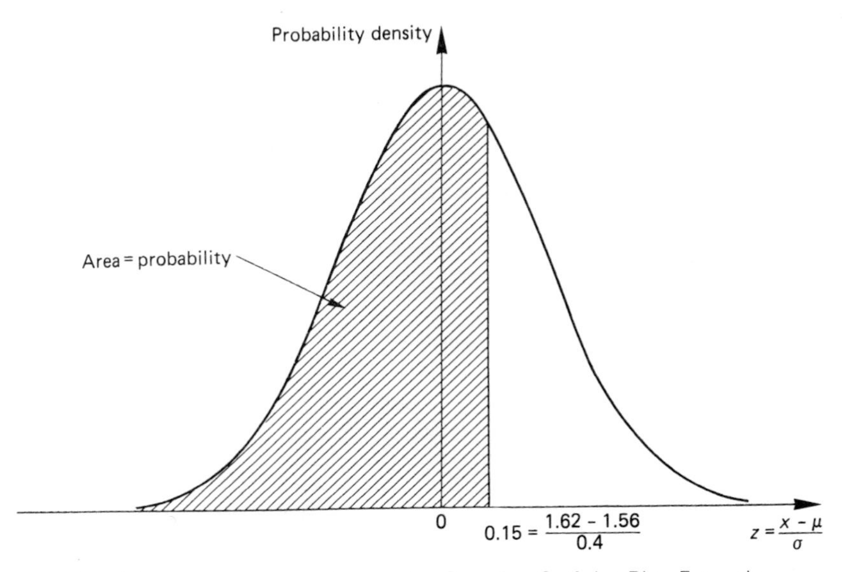


**Fig. 3.5** Standard normal curve for Question 2 of the Rice Example

0.15 is 0.440 38. As the total area under the standard normal curve is equal to 1, the shaded area in Fig. 3.5 is $1-0.440\,38$, which is 0.559 62. Hence **the probability of a bag chosen at random being less than 1.62 kg in weight is 0.56 or 56%**.

*Question 3* What is the probability that a bag chosen at random weighs between 1.6 kg and 1.75 kg?

*Solution* Proceeding straight to the standard normal curve diagram we get Fig. 3.6. The area indicated in this diagram can be thought of as being the area to the right of

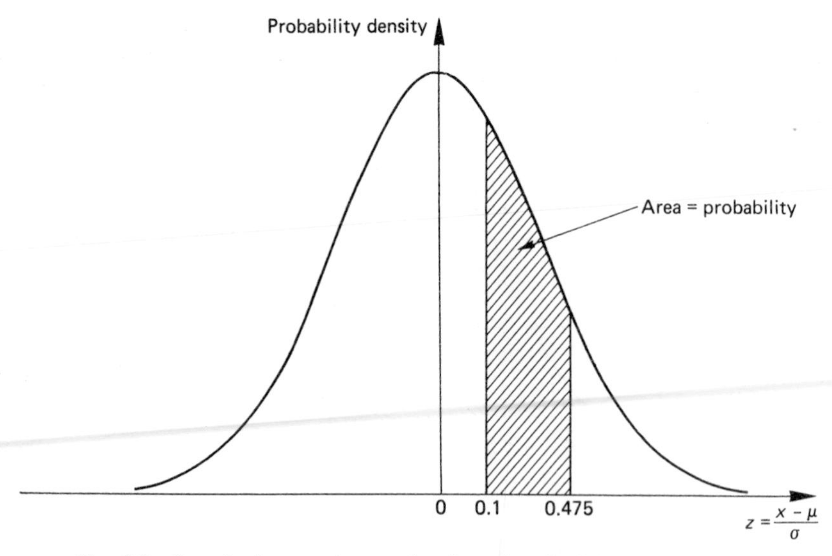


**Fig. 3.6** Standard normal curve for Question 3 of the Rice Example

0.1 excluding that to the right of 0.475. Both of these areas can be found from the tables and their values can be subtracted. Rounding 0.475 to 0.48 for this purpose, we obtain $0.46017 - 0.31561$, which is 0.14456. Therefore **the probability of choosing a bag of rice at random which weighs between 1.6 kg and 1.75 kg is 0.14**.

*Question 4* Find the probability that a bag chosen at random weighs more than 1.4 kg.

*Solution* The standard normal curve diagram is as in Fig. 3.7. When the $z$ score is negative the symmetry of the curve can be used to relate areas on either side of the

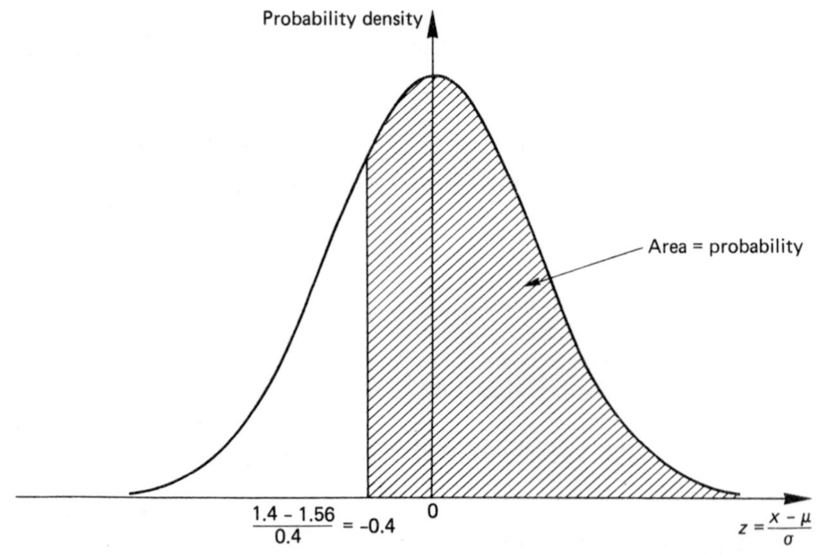


**Fig. 3.7** Standard normal curve for Question 4 of the Rice Example

vertical axis to each other. The area we want is 1 minus the area to the left of $-0.4$. This in turn is equal to the area to the *right* of $+0.4$, which from tables is 0.344 58. Hence the answer required is $1-0.344\,58$, that is 0.655 42. **The probability that a bag chosen at random weighs more than 1.4 kg is 0.66.**

The above questions have been concerned with the use of the tables on page 183 of areas under the standard normal curve. The ones on page 184 are commonly called **percentage points tables** as they enable us to look up the $z$ score corresponding to any given value for the area. The next question is an example of their use.

*Question 5* Within what range of values will 95% of the bags' weights lie?

*Solution* From the percentage points tables on page 184 we see that the central 95% of all $z$ scores are in the range $-1.96$ to $+1.96$, as in Fig. 3.8. This is because the $z$ score with area to the right equal to 0.025 is 1.96 from the tables.

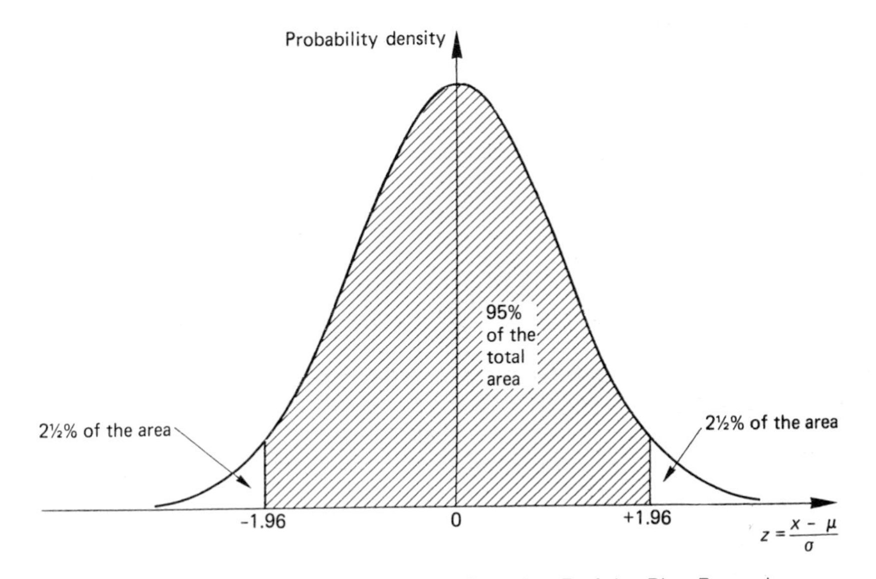


**Fig. 3.8** Standard normal curve for Question 5 of the Rice Example

Transposing equation (3.1) to make $x$ the subject gives a formula for $x$ in terms of $z$:

$$x = \mu + \sigma \,.\, z \tag{3.2}$$

Thus the extreme $z$ scores $-1.96$ and $+1.96$ correspond to the $x$ values

$$1.56-(0.4)\times(1.96) \qquad \text{and} \qquad 1.56+(0.4)\times(1.96) \tag{3.3}$$

which are 0.776 and 2.344. **We are 95% certain that a bag of rice chosen at random will weigh between 0.776 kg and 2.344 kg.** This is called a **95% prediction interval** for $x$.

*Question 6* The bags of rice are packed into boxes which contain 20 bags each. Ignoring the weight of the box itself, what is the probability that a full box weighs more than 33 kg?

*Solution* The normal distribution is **additive** in the sense that if two or more random variables are independent of each other and have normal distributions then their sum also has the normal distribution. The individual variables need not all have the same means and variances and the result is in fact true for any linear combination of them. Now the mean value of the sum of a number of random variables is equal to the sum of their mean values. Furthermore, the variance of a sum of independent variables is equal to the sum of their individual variances, so we can add the means and variances of any number of normally distributed independent random variables to derive the distribution of their total.

The boxes that form the population for this part of the example therefore have average weight $20 \times 1.56$ kg and variance $20 \times (0.4)^2$. The standard deviation has been squared here to give the variance and we shall need to take the square root of the resulting variance to find the standard deviation of the boxes' weight. Performing these calculations we find that the mean is 31.2 kg and the standard deviation is 1.7889 kg. The $z$ score for the $x$ value 33 is 1.01 and the area to the right of this under the standard normal curve is 0.15625. **The required probability that a box weighs more than 33 kg is 0.16 or 16%.**

## Classical sampling theory

For reasons given in Chapter 1, the means of large samples are better estimates of the population mean than those of smaller ones. This is known as the **Law of Large Numbers**. It can be shown that the relationship between sample mean variability, population variability and sample size is

$$\text{standard deviation of } \bar{x} = \frac{\text{standard deviation of } x}{\sqrt{(\text{sample size})}} \tag{3.4}$$

The right-hand side of this equation is called the **standard error in the mean** for a sample. It measures the variability between the averages of different samples whereas the standard deviation on the numerator describes the variability of the quantity $x$ within the population. It is by the act of sampling that we reduce the variability in the population to manageable proportions. The larger the sample size the smaller the discrepancies between different sample averages. This is the mathematical justification for the Law of Large Numbers.

There is an amazing and important result from the theory of statistics which extends the Law of Large Numbers described above and at the same time allows us to ignore to a certain degree the distribution of the parent population. The **central limit theorem** states that for a wide variety of populations the mean of a large sample is approximately normally distributed. The approximation becomes better as the sample size becomes bigger and provides a way of describing sample behaviour in order to utilise the reduction in variability that equation (3.4) expresses.

The existence of the Central Limit Theorem has led statisticians to adopt the following argument. A random variable either has the normal probability distribution or it does not have it. If it does, then the additivity property of that distribution implies that the mean of a random sample is itself normally distributed. If the random variable does not have the normal distribution, then the mean of a large random sample is roughly normally distributed anyway because of the Central Limit Theorem. In either eventuality, we may well assume for practical purposes that the parent population is normally distributed. It is in this fashion that the theorem seems to single out the normal distribution as being something special. The body of knowledge which forms the consequence of this **normality assumption** is called **classical sampling theory**. The Concrete Example of Chapter 10 contains a method for deciding whether a sample comes from a normal population or not.

The probability model that $x$ is normally distributed with mean $\mu$ and standard deviation $\sigma$ implies via additivity and equation (3.4) that the mean $\bar{x}$ of a sample of size $n$ is also normally distributed but with mean $\mu$ and standard deviation $\sigma/\sqrt{n}$. It is emphasised that we are now modelling the population of sample means rather than the parent population of individual $x$ values. Under the normality assumption it is the statistic

$$z = \frac{\bar{x} - \mu}{\sigma/\sqrt{n}} \tag{3.5}$$

which has the standard normal distribution. In practice the value of $\sigma$ in this equation has to be estimated by the sample standard deviation $s$ and the right-hand side then has the $t$ distribution. Before studying that and the behaviour of the sample variance we look at a slightly artificial situation in which equation (3.5) can be utilised directly.

## Sampling with known population standard deviation

The general procedure in analysing a sample of metric data is to derive estimates for both the population mean and the population standard deviation from it. Occasionally the population standard deviation is assumed to be known for certain and is not estimated. Possibly the sample is very small in size and an existing estimate is to be preferred. Here is an example in which the researcher does not have access to the raw data but knows only the average of the sample.

### The crisps example

Alf and Bob are executives who work for two rival potato crisp manufacturers. At lunch one day, Bob told Alf about a marketing research exercise for which a sample of 38 people were each given a packet of crisps to eat. He said that the average time a person took to consume a packet was 8.3 minutes. Much to Alf's disappointment, Bob could not tell him the standard deviation of the data. This was not surprising as they had both drunk several large whiskies.

*Question 1* What conclusion can Alf draw from the information he has?

*Solution* The sample average, 8.3 minutes, is a **point estimate** of the population mean. **Alf can infer that the population average time to eat a bag of crisps is about 8.3 minutes**.

**Point estimation** is not entirely satisfactory as it does not take into account the size of the sample nor the information on population variability that it contains. The same estimate could have been based on just two observations, one equal to 3.3 minutes and the other equal to 13.3 minutes. In that case the answer would not be very reliable as a description of population behaviour.

*Question 2* Back at his office the next day, Alf looks up his own company's marketing research information and finds from previous surveys that the standard deviation for the time to eat a packet of crisps is usually around 1.2 minutes. How does this knowledge of the population standard deviation help him to improve his estimate?

*Solution* We can quantify our confidence in an estimate by considering the probability distribution of the sample statistics involved. Under the normality assumption the $z$ score (3.5) has the standard normal distribution. By the same reasoning used in Question 5 of the Rice Example 95% of all samples have a $z$ value between $-1.96$ and $+1.96$. This can be expressed as a pair of inequalities:

$$-1.96 < \frac{\bar{x} - \mu}{\sigma/\sqrt{n}} < +1.96 \tag{3.6}$$

We now adopt the view that our sample does indeed fall into this central, 'middle-of-the-road' category. The population mean $\mu$ can be made the subject of (3.6) and so must satisfy

$$\bar{x} - \frac{1.96\sigma}{\sqrt{n}} < \mu < \bar{x} + \frac{1.96\sigma}{\sqrt{n}} \tag{3.7}$$

For the values of $\bar{x}$, $\sigma$ and $n$ appropriate to the crisps data, namely 8.3, 1.2 and 38 respectively we obtain the limits 7.92 and 8.68 from these inequalities. This is a 95% **confidence interval** for the population mean. **Alf can be 95% certain that the average time people take to eat a packet of crisps is between 7.92 minutes and 8.68 minutes.**

The general formula for a confidence interval for the mean when the population standard deviation is known is

$$\bar{x} - \frac{z\sigma}{\sqrt{n}} < \mu < \bar{x} + \frac{z\sigma}{\sqrt{n}} \tag{3.8}$$

The $z$ value is obtained from normal percentage points tables as in the Rice Example.

A range of values within which we have a specified degree of confidence a population parameter lies is called an **interval estimate** of the parameter. The point estimate, $\bar{x}$, obtained earlier is in the middle of confidence intervals given by the inequalities (3.8). The term added to and subtracted from $\bar{x}$ to form the extremities of a range can be thought of as the amount of statistical error in the sampling process.

*Question 3* Alf wants to be more than 95% certain of the range of possible population means. Calculate a 99% confidence interval for him.

*Solution* The central 99% of the area under the standard normal curve has $\frac{1}{2}$% in each of the two 'tails' on either side. The percentage points table on page 184 show that the $z$ score 2.57583 has $\frac{1}{2}$% of the area to its right and this is the $z$ value to be used in (3.8). The computation gives the result that **Alf can be 99% certain that the population mean is between 7.8 minutes and 8.8 minutes.**

It should be noted that this is a bigger range than the 95% confidence interval. The width of the range represents a compromise between how confident we want to be that it includes the true population mean and how small it must be to give worthwhile information about the population. After all, we can be 100% certain that the true population mean is between plus infinity and minus infinity but this information is hardly useful.

There is often misunderstanding about the interpretation of a confidence interval. The population mean is not a random variable but a fixed and precise mathematical quantity. The inference about its value consists of a statement about the behaviour of 95% or 99% or whatever of all possible samples. We are saying that the most likely $q$% of all possible samples will contain the true population mean within their $q$%

confidence interval. For this reason we are $q\%$ certain that the true value is inside the range we calculate.

## Sampling with unknown population standard deviation

The Crisps Example above is unusual in that the standard deviation of the sample is not available to us. It is more likely that we can calculate the sample standard deviation along with its average to give an estimate of the population standard deviation $\sigma$. Unfortunately, replacing $\sigma$ by $s$ in equation (3.5) alters its probability distribution and it is no longer a standard normal variable.

## The $t$ and $\chi^2$ distributions

The statistic

$$t = \frac{\bar{x} - \mu}{s/\sqrt{n}} \tag{3.9}$$

has the $t$ distribution with $(n-1)$ degrees of freedom. The concept of degrees of freedom was discussed briefly in the Shower Example of Chapter 2. All it means here is that the exact nature of the $t$ distribution depends on the size of the sample being used.

The statistic

$$\chi^2 = \frac{(n-1)s^2}{\sigma^2} \tag{3.10}$$

has the $\chi^2$ distribution, again with $(n-1)$ degrees of freedom. The Greek letter $\chi$ is pronounced 'ki' to rhyme with 'pie' and the number of degrees of freedom is usually denoted by another Greek letter, $\nu$, pronounced 'nu'. The next example illustrates the use of these probability distributions in the analysis of a sample.

### The train journey example

A commuter timed his train journey to the office on 8 separate occasions. The results, measured in minutes, were 25.4, 27.0, 29.3, 24.4, 25.9, 27.4, 25.9 and 28.0.

*Question 1* Determine a 95% confidence interval for the average time the train journey takes.

*Solution* The sum of the data values is 213.3 and the sum of their squares is 5704.39. Formulae (2.1), (2.3) and (2.7) for the mean, variance and standard deviation of a sample give the following:

$$x = \frac{213.3}{8} = 26.6625$$

$$s^2 = \frac{5704.39 - ((213.3)^2/8)}{(8-1)} = 2.4684 \tag{3.11}$$

$$s = \sqrt{2.4684} = 1.571$$

We could consider these sample statistics as point estimates of the corresponding population parameters. For reasons given in the previous example it is often preferable to quote an interval estimate, and the $t$ distribution provides a pair of inequalities similar to (3.8):

$$\bar{x} - \frac{ts}{\sqrt{n}} < \mu < \bar{x} + \frac{ts}{\sqrt{n}} \tag{3.12}$$

Tables of the precentage points of the $t$ distribution can be found on page 192. Our $t$ statistic has 7 degrees of freedom and so from those tables the $t$ value 2.365 has $2\frac{1}{2}\%$ of the area under the $t$ distribution curve to its right, as in Fig. 3.9. This diagram should be compared with Fig. 3.8 for the normal curve. Although it has the same bell shape in our sketches, the $t$ distribution is slightly flatter.

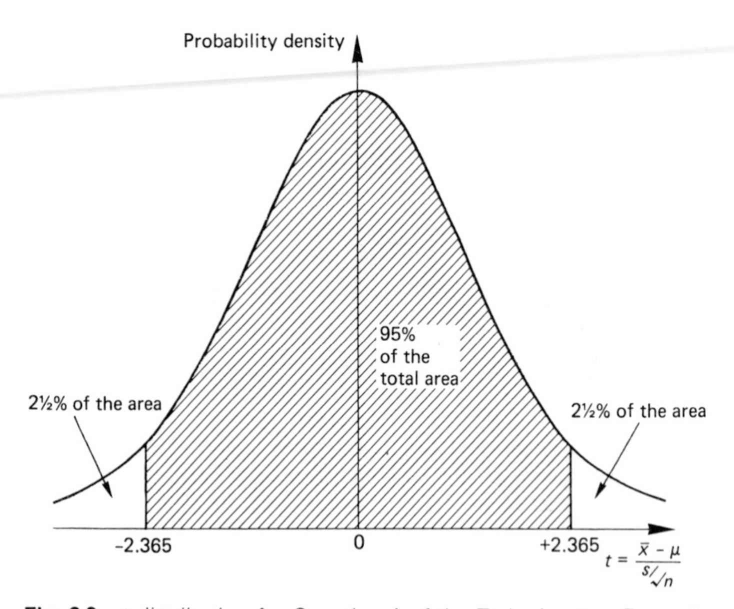


**Fig. 3.9** *t* distribution for Question 1 of the Train Journey Example

For large samples the value of $s$ is a good estimate of the population standard deviation $\sigma$ and so the $t$ statistic (3.9) is almost the same as the normal statistic (3.5). It follows that the distributions are almost the same and for more than 50 degrees of freedom the $t$ statistic can be assumed to be normally distributed.

Substituting $\bar{x}$, $t$, $s$ and $n$ as 26.6625, 2.365, 1.571 and 8 into (3.12) we find that **the 95% confidence interval for the average journey time is between 25.3 minutes and 28.0 minutes.**

*Question 2* Before he collected the data, the commuter believed that the average journey time was 25 minutes. Does the answer to Question 1 contradict this belief?

*Solution* The statement that the population mean is 25 minutes is a **null hypothesis, $H_0$**. It embodies a preconception we had about the population before the data was collected and which we seek evidence to disprove. The argument that disproof is an easier procedure than proof was given in the section on scientific method in Chapter 1. The sample of just 8 journey times we have here cannot be expected to prove conclusively that the population mean of all journey times is exactly equal to a particular value. However, it can enable us to assess in probabilistic terms whether a specific null hypothesis is plausible or not.

From the answer to Question 1, we are 95% certain that the population mean is in the range 25.3–28.0. The null hypothesis value, 25, is outside this range. Either we have chosen one of the rare, 5% likely samples whose 95% confidence interval does not contain the population mean or the mean is not 25. Acknowledging that there is a 5% chance we are wrong, we reject the null hypothesis and state that **the population mean is not 25 minutes.**

The probability level of 5% at which we have tested the null hypothesis is called the **level of significance** of the test. Rejecting a null hypothesis at this level is equivalent to saying that we are 95% confident that it is false.

There is a more revealing way of testing a null hypothesis than using confidence intervals for the parameter in question. Instead of forming such an interval from the $t$ statistic (3.9) we can calculate its value assuming the null hypothesis is correct. This gives a **test statistic**:

$$t = \frac{\bar{x} - \mu}{s/\sqrt{n}} = \frac{26.6625 - 25}{1.571/\sqrt{8}} = 2.99 \tag{3.13}$$

The number of degrees of freedom to be associated with this statistic is $(8-1)$, as before. Now if the null hypothesis is true, then the $t$ statistic will be relatively small as $\bar{x}$ should be near to 25. If $t$ is large then there may be evidence that it is false.

From the tables on page 192 we see that for 7 degrees of freedom, $2\frac{1}{2}$% of the area under the $t$ curve is to the right of 2.365. This is precisely the same $t$ value we used in the solution to Question 1 and the appropriate diagram is Fig. 3.9. The 5% level of significance we are using has the effect of dividing the range of all possible $t$ values into an **acceptance region**, $-2.365$ to $+2.365$, with a high probability of occurring by chance, and a **rejection region**, values outside this range, which has a relatively low probability. The boundaries between these regions are marked by **critical values** of the test statistic, in this case $\pm 2.365$. The whereabouts of our test statistic determines whether we accept or reject the null hypothesis. Figure 3.9 shows that the rejection region consists of the two tails of the distribution. The $t$ statistic given by (3.13) falls in this region and so, again, **we reject the null hypothesis that the population mean is 25 minutes.** We have carried out a **two-sided** or **two-tail** test and would have rejected the null hypothesis if the $t$ statistic had been either large and negative or large and positive. In the solution to the next question it is necessary to perform a one-tail test.

*Question 3* Is there evidence that the average journey time is significantly greater than 24.5 minutes?

*Solution* In the previous question we investigated whether there was evidence that $\mu$ is not equal to 25. We would have rejected the hypothesis that it is 25 if the sample estimate $\bar{x}$ were either significantly bigger or smaller than 25. The two possibilities for rejection implied that we performed a two-tail test of the $t$ statistic. Question 3, on the other hand, is concerned not with whether $\mu$ is unequal to 24.5 but whether it is greater than 24.5. This is a one-sided approach and we test $t$ for being large and positive as opposed to being small or negative. In other words we want evidence that the sample mean $\bar{x}$ is significantly bigger than 24.5 rather than just being different from 24.5.

According to the percentage points table for the $t$ distribution on page 192, 5% of the area under the $t$ curve with 7 degrees of freedom is to the right of 1.895. This is

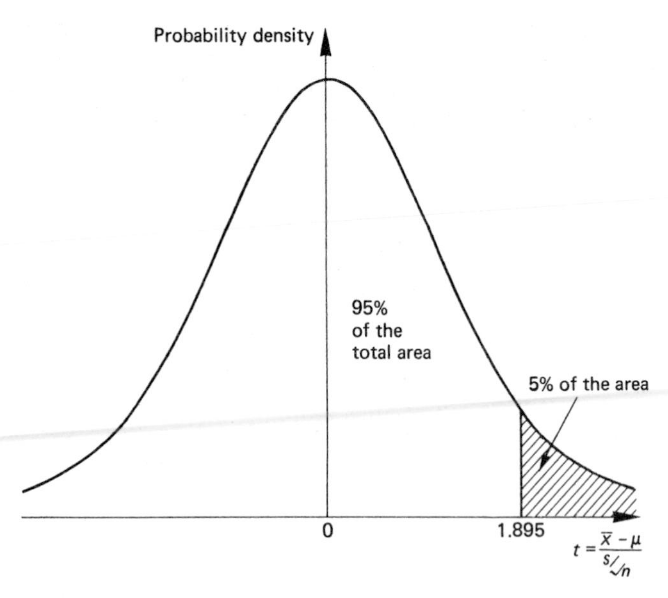


**Fig. 3.10** *t* distribution for Question 3 of the Train Journey Example

illustrated in Fig. 3.10. The test statistic is:

$$t = \frac{\bar{x} - \mu}{s/\sqrt{n}} = \frac{26.6625 - 24.5}{1.571/\sqrt{8}} = 3.89 \tag{3.14}$$

and it lies in the rejection region. Hence **there is evidence at the 5% level of significance that the population mean journey time is greater than 24.5 minutes.**

As the rejection region is on one side of the distribution curve the test is **one-sided** or **one-tail**. The null hypothesis is that $\mu$ is equal to 24.5 and forms a working hypothesis with which we can calculate a value for the test statistic in equation (3.14). The difference between a one-tail and a two-tail test can be expressed in terms of the **alternative hypothesis**. This is the hypothesis we accept when we reject the null hypothesis. For a two-tail test it is that $\mu$ is not equal to the hypothesised value. For the one-tail test we have applied it was that $\mu$ is greater than the null hypothesis value of 24.5. Other one-tail tests may seek to show that it is less than some given value. A one-tail test allows a researcher to determine whether a treatment produces a significant and positive improvement in some response variable rather than just a change in it.

The level of significance at which a test is carried out is the probability of rejecting a null hypothesis which is in fact correct. Doing this is known as making a **type I error**. A **type II error** is to accept a null hypothesis which is really false. It is unfortunate that the smaller the level of significance and hence the probability of a type I error, the larger the probability of making a type II error and vice versa. It is therefore considered prudent to start at the level of 5% and try to 'chase along' the significance in the tables. Looking across the row corresponding to 7 degrees of freedom for the $t$ distribution on page 192 we see that our $t$ value of 3.89 is larger than the critical value at 0.5%. The test statistic is thus significant at 0.5%, highly significant indeed. The conclusion can be reworded to say that **there is evidence at the 0.5% level of significance that the population mean is greater than 24.5 minutes.**

*Question 4* Determine a 95% confidence interval for the variance of the journey times.

*Solution* As stated earlier, the quantity $(n-1)s^2/\sigma^2$ has the $\chi^2$ distribution with $(n-1)$ degrees of freedom. Our sample size is 8 and according to the tables on page 188 $2\frac{1}{2}\%$ of the area under the $\chi^2$ distribution with 7 degrees of freedom lies to the left of 1.6899 and $2\frac{1}{2}\%$ to the right of 16.0128. Hence we can be 95% certain that

$$1.6899 < \frac{(n-1)s^2}{\sigma^2} < 16.0128 \tag{3.15}$$

Transposing these inequalities to make $\sigma^2$ the subject:

$$\frac{(n-1)s^2}{16.0128} < \sigma^2 < \frac{(n-1)s^2}{1.6899} \tag{3.16}$$

Substituting the values 8 and 2.4684 for $n$ and $s^2$ we find that **the 95% confidence interval for the population variance is 1.08 minutes$^2$ to 10.22 minutes$^2$.**

Interval estimates for the population variance like the one we have just calculated are not encountered very often. Usually the point estimate $s^2$ is adequate. The $\chi^2$ distribution, however, plays an important role in the next section, but in a different guise.

## Testing probability models

The 'goodness of fit' with which a set of expected frequencies describes a set of observed ones can be measured using the $\chi^2$ distribution. The first example raises the question of whether three attributes occur equally often within a certain population.

### The newspaper example

A person advertised an article for sale in three different newspapers. The replies were distributed as follows:

| Newspaper | *Daily Sheet* | *Morning Chorus* | *Dawn Patrol* |
|---|---|---|---|
| Number of replies | 53 | 69 | 43 |

*Question 1* Are the three newspapers equally effective in attracting responses to the advertisement or are the differences in the numbers of replies significant?

*Solution* There were 165 replies altogether. If the newspapers are equally effective, which is our null hypothesis, then there should have been (165/3) responses from each one. We must therefore decide whether the observed frequencies 53, 69, 43 are sufficiently close to 55, 55, 55 to support the null hypothesis or sufficiently different from 55, 55, 55 to warrant its rejection.

The $\chi^2$ statistic measures the overall difference between observed frequencies and values we expect to occur based on a null hypothesis. It is defined as:

$$\chi^2 = \text{sum of} \left\{ \frac{(\text{observed frequency} - \text{expected frequency})^2}{\text{expected frequency}} \right\} \tag{3.17}$$

The value of $\chi^2$ is a sort of total squared error of prediction from observation. If all the observed frequencies were equal to the expected ones then it would be zero. The greater the discrepancy between observation and expectation the larger the value of $\chi^2$ will be.

Using the data in the question and the predictions 55, 55, 55 in (3.17):

$$\chi^2 = \frac{(53-55)^2}{55} + \frac{(69-55)^2}{55} + \frac{(43-55)^2}{55}$$

$$= 0.0727 + 3.5636 + 2.6182 = 6.255 \qquad (3.18)$$

The tables on page 188 give a confidence interval for the $\chi^2$ statistic assuming that the null hypothesis is true and the observed frequencies are just random fluctuations from the expected ones. The tables take into account the number of frequencies being predicted via the degrees of freedom parameter of the distribution. Clearly the accumulated squared error of 6.255 would be more impressive if there had been, say, 50 forecasts rather than the three we actually did make. The number of degrees of freedom associated with a goodness of fit test is

$$\nu = (\text{number of frequencies being predicted})$$
$$-(\text{number of quantities calculated from the observations}$$
$$\text{in order to make the predictions}) \qquad (3.19)$$

If there is only 1 degree of freedom, then Yates' correction should be applied as described in the Coin Example later in this chapter. For the newspaper sample there are three frequencies being predicted with one calculation performed on the data, namely adding them up to obtain 165.

The null hypothesis for a $\chi^2$ goodness of fit test is that $\chi^2$ should be equal to zero. The larger the value we get, the more unlikely it is to be a random fluctuation from the 'ideal' of zero. The percentage points table for the $\chi^2$ distribution on page 188 shows that for two degrees of freedom 5% of the area under the density curve is to the right of 5.9915. We perform a one-tail test, as small values of $\chi^2$ indicate that the fit is good and we seek evidence to reject that possibility, which means we are testing for a significantly high value.

Our calculated $\chi^2$ is 6.255, which is larger than 5.9915 and, refusing to believe that our sample is so extreme as to be less than 5% likely to have occurred by chance, we reject the null hypothesis at the 5% level of significance. **The newspapers are significantly different from each other at the 5% level in the number of responses they attracted to the advertisement.**

It is desirable for each expected frequency in a $\chi^2$ test to be at least 5, although this is not absolutely necessary and some authorities maintain that if they are bigger than zero, then all will be well. Categories of data can be merged to give sufficiently large frequencies before the test is begun and this is illustrated in the Toaster and Caterpillar Examples later in this chapter. The overall sample size for the test should be at least 20 and preferably above 40. The references quoted in the section on Further Reading give more information on these points.

Taking the expected frequencies to be all the same was equivalent to assuming that each response had an equal probability of being from each of the three newspapers. When every outcome has the same chance of occurring in a trial then the probabilities have the **uniform** distribution. The continuous version of this distribution was used in the section on probability in Chapter 1.

*Question 2* It costs £4 to place an advertisement in the *Morning Chorus* and £2 each for the other two newspapers. Are the levels of response consistent with these differences in cost?

*Solution* The null hypothesis does not have to imply equal predicted frequencies and here we test whether the observations are in accordance with the costs after allowing for statistical error. The costs are in the ratio 2:4:2 and apportioning the observed total, 165, in this ratio gives expected frequencies of 41.25, 82.50, and 41.25. The $\chi^2$ statistic is:

$$\chi^2 = \frac{(53-41.25)^2}{41.25} + \frac{(69-82.50)^2}{82.50} + \frac{(43-41.25)^2}{41.25} = 5.63 \tag{3.20}$$

We have calculated one quantity from the observed frequencies, their total, and so, according to equation (3.19) the degrees of freedom to be associated with this statistic is 2. The critical value of $\chi^2$ at 5% for two degrees of freedom from the tables on page 188 is 5.9915. Hence we conclude that as 5.63 is not bigger than 5.9915 we cannot reject the null hypothesis. **The frequencies of the responses could well be explained by the differences in circulation figures reflected in the differences in costs of the advertisements in the three newspapers.** Note that we have not proved the null hypothesis. No amount of sampling can prove conclusively that a generalisation is absolutely true. What is stated here is that we have failed to disprove it.

## Yates' correction

When the number of degrees of freedom given by equation (3.19) is equal to 1, a correction due to Yates should be applied. Each difference between the observed and expected frequencies is reduced in size by 0.5 before being squared in formula (3.17) for $\chi^2$. Here is an example.

### The coin example

A coin is tossed 131 times and lands 'heads' on 73 occasions. Is this evidence that the coin is not a fair one?

*Solution* The null hypothesis that the coin is fair leads to the prediction that half the tosses should be 'heads' and half 'tails' (Table 3.1).

**Table 3.1** Calculations for the Coin Example

| Outcome | Heads | Tails |
|---|---|---|
| Observed frequency | 73 | 131 − 73=58 |
| Expected frequency | 131/2=65.5 | 131/2=65.5 |
| Observed frequency − expected frequency | +7.5 | −7.5 |

There are two predicted frequencies and by adding the observations together to make the predictions we have performed one calculation on the raw data. Hence there are (2 − 1) degrees of freedom from equation (3.19). As this equals 1 we apply Yates' correction to equation (3.17):

$$\chi^2 = \frac{(7.5-0.5)^2}{65.5} + \frac{(7.5-0.5)^2}{65.5} = 1.496 \tag{3.21}$$

Note that the numerical size of the differences in Table 3.1 are reduced by 0.5. This will be a subtraction of 0.5 if the observed frequency is bigger than the expected one or an addition of 0.5 if it is not.

The tables on page 188 indicate that $\chi^2$ has to reach 3.8415 when $\nu$ is 1 before it is significantly large at the 5% level. Our value does not do this and so we cannot reject the null hypothesis. There is no evidence that the frequencies are significantly different from 65.5 and hence **no evidence that the coin is unfair.**

The $\chi^2$ goodness of fit test is extremely useful. We now apply it to test the suitability of three important probability models to specimen sets of data.

## The binomial distribution

The binomial distribution formula gives the probability of exactly $r$ successes when $n$ identical trials are performed. Each trial must have the same probability $p$ of being successful and not be influenced by the outcomes of the other trials.

$$P(r \text{ successes out of } n \text{ trials}) = \frac{n!}{(n-r)!r!} p^r (1-p)^{n-r} \qquad (3.22)$$

The exclamation mark ! following a number $n$ means multiply all the numbers from 1 up to $n$ together. For instance 4! stands for $1 \times 2 \times 3 \times 4$ which is 24. It is read '$n$ factorial' and is found as a function on most scientific calculators. By convention, 0! is taken to be equal to 1.

The **binomial distribution** is the set of $(n+1)$ probabilities $P(0)$, $P(1)$, $P(2)$, . . . , $P(n)$ obtained by putting $r$ equal to each number from 0 to $n$ in equation (3.22). To help the reader gain familiarity with the formula, we now do a simple calculation with it.

Suppose a gambler has a 24% chance of winning a card game. He wants to know the probability of winning exactly 4 out of a sequence of 9 games he intends to play. Each game is a separate trial and the values of $n$ and $p$ are 9 and 0.24. Equation (3.22) becomes, with $r$ equal to 4:

$$P(4 \text{ successes out of } 9 \text{ trials}) = \frac{9!}{(9-4)!4!}(0.24)^4 \times (1-0.24)^{9-4}$$

$$= \frac{362\,880}{120 \times 24} \times 0.00332 \times 0.25355 = 0.106 \qquad (3.23)$$

Hence the gambler has a probability of about 11% of winning exactly 4 of the 9 games he plays.

### The toaster example

The manager of a manufacturer's service department collects the following data on 3248 electric toasters:

| Number of defective elements in first year of life | 0 | 1 | 2 | 3 | 4 |
|---|---|---|---|---|---|
| Number of toasters | 2286 | 836 | 115 | 9 | 2 |

It has always been assumed that each of the 4 heating elements in the toaster has the same probability of failure irrespective of its position and will fail independently of the others. Does the data support these assumptions?

*Solution* The trial 'has the element failed' is being repeated 4 times for each toaster during the first year of its life. The data represents the observed number of occurrences of the outcomes 0, 1, 2, 3 and 4 failures for the sequence of 4 such trials. For instance there were 115 occasions when exactly 2 out of the 4 trials were 'success' in that the element failed.

Under the null hypothesis that the probability of an individual element failing is the same for all of them and they fail independently, the binomial distribution should describe the frequencies which have been observed. Table 3.2 shows how $p$ is calculated from the observed frequencies.

**Table 3.2** Estimation of $p$ for the Toaster Example

| Number of defective elements in toaster ($r$) | 0 | 1 | 2 | 3 | 4 | Total |
|---|---|---|---|---|---|---|
| Observed frequency ($f$) | 2286 | 836 | 115 | 9 | 2 | 3248 |
| Total number of defective elements ($f \times r$) | 0 | 836 | 230 | 27 | 8 | 1101 |

$$p = P(\text{individual element fails}) = \frac{\text{number of failed elements}}{\text{total number of elements}}$$

$$= \frac{\text{number of failed elements}}{4 \times \text{total number of toasters}} = \frac{1101}{4 \times 3248} = 0.0847$$

Putting $n$ and $p$ equal to 4 and 0.0847 in equation (3.22) gives the required probability distribution (Table 3.3).

Notice that the smallest expected frequency is below the desirable minimum of 5 for the $\chi^2$ test and so has been merged with that of the adjoining category. There are now 4 predicted frequencies and we have performed two calculations on the data, adding the observations together and also determining the total number of failed elements. According to (3.19) the number of degrees of freedom is therefore $(4-2)$ for the $\chi^2$ statistic and as this is bigger than 1 Yates' correction is unnecessary. The 5% critical value for $\chi^2$ with 2 degrees of freedom is 5.9915 from the tables and so

**Table 3.3** The binomial distribution for the Toaster Example

| Number of defective elements ($r$) | 0 | 1 | 2 | 3 | 4 |
|---|---|---|---|---|---|
| Probability based on the binomial distribution | 0.7019 | 0.2598 | 0.0361 | 0.0022 | 0.0001 |
| Expected frequency of toasters = probability × 3248 | 2279.8 | 843.8 | 117.1 | 7.1 | 0.3 |
| | | | | 7.4 (merged) | |
| Observed frequency | 2286 | 836 | 115 | 9 | 2 |
| | | | | 11 (merged) | |
| Contribution to $\chi^2$ | 0.017 | 0.072 | 0.038 | 1.751 | |

Total $\chi^2 = 1.88$

the observations are not outside the limits of being random fluctuations from binomial frequencies. **The data could have arisen from a sample of toasters whose elements were failing independently of each other with the same probability, 0.0847.**

In some applications of the binomial distribution the value of the probability of success on each trial, $p$, is part of the null hypothesis. For example in testing a die we might assume that throwing a '3' has probability $\frac{1}{6}$. In that case the data would be used only once in determining the expected frequencies and the number of degrees of freedom would be just one less than the number of predictions.

## The Poisson distribution

The Poisson distribution describes events which occur randomly in time or space. For example, flashes of lightning are random events in time and flaws in fabric are random in space. If there are on average $a$ occurrences of the event in question within a certain time period or spatial region, then

$$P(r \text{ occurrences}) = \frac{e^{-a}a^r}{r!} \tag{3.24}$$

The letter e in this formula is a universal constant, like $\pi$, whose value is approximately 2.7183. Scientific calculators have the exponential function $e^x$ built into them. Here is an example on the fitting of a Poisson distribution to a sample.

### The caterpillar example

A biologist counts the numbers of caterpillars on 299 leaves:

| Number of caterpillars | 0 | 1 | 2 | 3 | 4 | 5 | 6 or more |
|---|---|---|---|---|---|---|---|
| Number of leaves | 62 | 74 | 68 | 53 | 22 | 14 | 6 |

If the leaves are all roughly the same size and the caterpillars wander randomly, then the Poisson distribution should describe these frequencies. We investigate whether this is so.

*Solution* It is necessary to estimate the average number of caterpillars on a leaf before the Poisson formula can be used. A decision must be made on exactly how many

**Table 3.4** Estimation of $a$ for the Caterpillar Example

| Number of caterpillars per leaf ($r$) | 0 | 1 | 2 | 3 | 4 | 5 | 7 | Total |
|---|---|---|---|---|---|---|---|---|
| Number of leaves ($f$) | 62 | 74 | 68 | 53 | 22 | 14 | 6 | 299 |
| Total number of caterpillars ($f \times r$) | 0 | 74 | 136 | 159 | 88 | 70 | 42 | 569 |

$$a = \frac{\text{number of caterpillars}}{\text{number of leaves}} = \frac{569}{299} = 1.903$$

caterpillars the category '6 or more' should count as in this calculation. We assume that the average for the 6 leaves which fell into this category was 7 insects although this figure would usually be suggested by the researcher. The estimation is shown in Table 3.4.

The Poisson formula (3.24) gives the probabilities of the various possible numbers of caterpillars on a leaf. Multiplying these by the total number of leaves, 299, gives the frequencies to be expected if the null hypothesis that the Poisson distribution is appropriate is true. Table 3.5 displays the calculations together with the observed frequencies from the question. The probability of the outcome '6 or more' is found by subtracting all the other probabilities from 1. The fraction of leaves having 6 or more caterpillars on them is 1 minus the fraction with 5 or less.

**Table 3.5** The Poisson distribution for the Caterpillar Example

| Number of caterpillars ($r$) | 0 | 1 | 2 | 3 | 4 | 5 | 6 or more |
|---|---|---|---|---|---|---|---|
| Probability based on the Poisson distribution | 0.149 | 0.284 | 0.270 | 0.171 | 0.081 | 0.031 | 0.014 |
| Expected frequency of leaves = probability × 299 | 44.6 | 84.9 | 80.7 | 51.1 | 24.2 | 9.3 | 4.2 |
| | | | | | | 13.5 | |
| Observed frequency | 62 | 74 | 68 | 53 | 22 | 14 | 6 |
| | | | | | | 20 | |
| Contribution to $\chi^2$ | 6.79 | 1.40 | 2.00 | 0.07 | 0.20 | 3.13 | |

Total $\chi^2 = 13.59$

As the observations and expectations do not seem close to each other the contribution of each prediction to the $\chi^2$ statistic in equation 3.17 is recorded in the calculation. We shall discuss this shortly. Apart from that row, the working is similar to that for the last example and again the smallest two expected frequencies have been merged to avoid any of them being less than 5 in value.

As for the binomial, the number of degrees of freedom for the $\chi^2$ statistic when fitting a Poisson distribution is 2 less than the number of predicted frequencies. This is because the data has been totalled and an estimate of the average outcome taken from it. Our $\chi^2$ statistic thus has $(6-2)$ degrees of freedom. The 5% critical value for 4 degrees of freedom is 9.4877 and so the calculated value is significantly large. It can be seen from the tables that our $\chi^2$ is significant at the 1% level and so has a less than 1% chance of being a random fluctuation from null hypothesis behaviour. **We therefore reject the null hypothesis that the observed frequencies can be predicted with the aid of the Poisson distribution at the 1% level of significance.**

In the last two examples the expected frequencies were satisfactory descriptors of the observed ones and the value of $\chi^2$ was small. It was not necessary to take the analysis any further in those cases as the model adopted seemed to be adequate. In this example we reject the predictions given by the null hypothesis distribution and it is instructive to investigate possible reasons for the lack of fit.

The assumptions implicit in the use of the Poisson distribution are that occurrences of the event concerned are random and independent of each other. If either of these conditions is not met then the Poisson model may fail. The components of $\chi^2$ in Table 3.5 provide clues as to the whereabouts of the discrepancies between prediction and observation.

The largest contributions, and therefore the largest errors, are from the categories '0 caterpillars on a leaf' and '5 or more caterpillars on a leaf'. The insects do not seem to treat vacant or full leaves in a completely random manner. Possibly there are factors which preclude certain leaves from hosting caterpillars altogether. They may be diseased, awkardly positioned or in unsuitable light. Leaves with lots of caterpillars on them may become 'saturated' and so the chance of more arrivals to them is small.

Our conclusion suggests that a more detailed experiment should be performed, examining unoccupied and overcrowded leaves in particular, to explain the caterpillars' non-random behaviour.

## The Poisson approximation to the binomial distribution

The Poisson formula (3.24) can be used to approximate the results of the binomial formula (3.22) in certain circumstances. When the number of trials, $n$, is large and $p$, the probability of success at each trial, is small, the binomial formula becomes very difficult to work with. The juxtaposition of large factorials and powers of numbers less than 1 which are very small mean that the number range in the calculation is too great for calculators and computers alike.

The approximation consists of using the Poisson with the value of $a$ set equal to $pn$. It must be emphasised that the resulting probabilities will be valid only if the binomial formula itself cannot be computed because of numerical problems.

## Normal approximations to the Poisson and binomial distributions

The normal distribution provides an approximation to both the binomial and the Poisson models when their formulae become difficult to evaluate because of the size of the numbers being used. The Prime Minister Example later in the chapter involves the normal approximation to the binomial distribution while the next example is of an approximation to the Poisson.

### The indigestion example

A hospital specialist finds that his patients report an average of 153.8 attacks of indigestion each per year. If the attacks are occurring randomly in time, what is the probability that a particular patient suffers between 162 and 176 attacks inclusive in a given year?

*Solution* The assumption that the attacks occur randomly in time allows us to use the Poisson distribution as a model of their frequency. In the Caterpillar Example the distribution described spatial randomness but its main applications are to events which are random in time. The average $a$ for the patient with indigestion is 153.8 and we would like to substitute this, together with the $r$ values 162, 163, 164, 165, . . . , 175 and 176, in equation (3.24). The calculations are difficult if not impossible because of the astronomical size of the factorials which arise.

When the Poisson formula cannot be utilised then the normal distribution with mean $a$ and standard deviation $\sqrt{a}$ gives a good approximation. A **correction for continuity**, similar to Yates' correction for $\chi^2$, has to be incorporated. The normal distribution relates to a random variable $x$ which is continuous, whereas the binomial and Poisson describe outcomes which must be whole numbers. It is therefore necessary to make the correspondence between the $x$ axis of the normal curve and the whole numbers of the distribution being approximated very carefully.

The logical way of establishing this correspondence is to represent each whole number by the set of those $x$ values which round off to that number as their nearest integer. Hence the range '162 to 176 attacks inclusive' in this example corresponds to the range '$x$ is between 161.5 and 176.5' on the normal curve $x$ axis, because every member of this range, when rounded off to the nearest whole number, is between 162 and 176 inclusive. The standard normal curve diagram is shown in Fig. 3.11. Subtracting the relevant areas to the right of the $z$ scores we find that **the probability of a patient suffering between 162 and 176 attacks inclusive during a particular year is 0.23.**

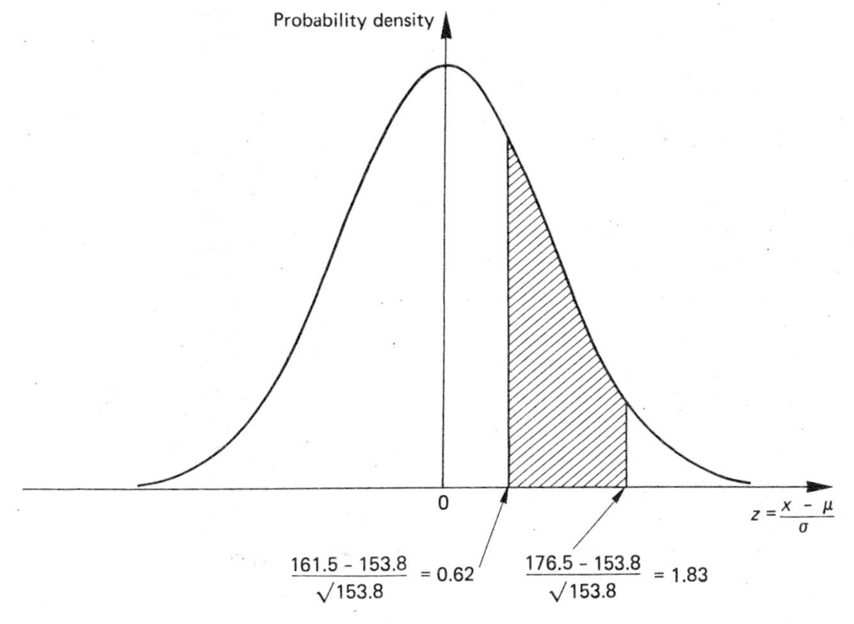


**Fig. 3.11** Standard normal curve for the Indigestion Example

## Population proportions

The fraction of times a particular attribute occurs within a population is often of interest. The normal approximation to the binomial distribution can be used to describe the behaviour of a percentage or a proportion. Alternatively the frequencies which occurred in the sample can be tested with the $\chi^2$ statistic as in the Newspapers Example. The equivalence of performing a $\chi^2$ test on two expected frequencies to the normal approximation to the binomial is discussed later in the chapter.

### The Prime Minister example

In a sample of 534 respondents, 246 answered 'Yes' to the question 'Do you think that the Prime Minister is giving the country effective leadership?' Estimate the percentage of people in the population who support the Prime Minister.

*Solution* We can obtain a point estimate of the population proportion by simply dividing 246 by 534. The result, 0.4607 or 46%, does not take the size of the sample into account and could have been based on a survey of just 100 people. As in the Train Journey Example earlier, our confidence in the estimate can be quantified by considering the probability distribution of the statistic concerned to produce an interval estimate.

A 'Yes' response to the question can be thought of as being a 'success' in a binomial trial. Suppose the probability is $p$ that a person chosen at random says 'Yes'. It follows that the probability of a specific number of successes in our sample is given by formula (3.22) with $n$ equal to 534. When the value of $n$ is as large as this the factorials in the binomial formula are difficult to compute. The Poisson distribution with mean $pn$ can be utilised in such cases but the numbers of successes in this example are in the hundreds and are too big for that formula as well.

If the original binomial equation and the Poisson approximation both fail, then the numbers are large enough for the normal distribution to be a valid substitute. Hence the number of successes is normally distributed with mean $pn$ and standard deviation $\sqrt{[np(1-p)]}$. **It follows that the sample proportion of successes, $(x/n)$, is also normally distributed with mean $p$ and standard deviation $\sqrt{[p(1-p)/n]}$.**

The normal percentage points table on page 184 shows that $2\frac{1}{2}$% of the area under the standard normal curve lies to the right of 1.96. Hence the central 95% of all possible $z$ scores are between $-1.96$ and $+1.96$. A diagram to illustrate this is in Fig. 3.8 for the Rice Example. The extreme $z$ scores correspond to the $x$ values $\mu-1.96\sigma$ and $\mu+1.96\sigma$ from equation (3.2). Inserting the appropriate expressions for $\mu$ and $\sigma$ and dividing by $n$ the prediction interval for the fraction of successes is

$$p-1.96\sqrt{\frac{p(1-p)}{n}} < \text{observed fraction, } h < p+1.96\sqrt{\frac{p(1-p)}{n}} \qquad (3.25)$$

This tells us that 95% of all samples will have fractions of 'Yes' responses between the two limits stated. As in the calculation of a confidence interval for the mean in the Crisps Example, we now need to make the population parameter the subject of this pair of inequalities. It can be shown that to the degree of approximation to which we are working:

$$h-1.96\sqrt{\frac{h(1-h)}{n}} < p < h+1.96\sqrt{\frac{h(1-h)}{n}} \qquad (3.26)$$

Substituting the observed proportion, $h=0.4607$, for this example into (3.26) gives the result that we can be 95% certain that the proportion is between 0.4184 and 0.5030. We could quote this conclusion by saying that we are 95% certain that **the proportion of people who support the Prime Minister is between 42% and 50%**. As for other confidence intervals, the corresponding point estimate of 46% is in the middle of the range.

The general formula for the extremes of the confidence interval of a proportion is an obvious consequence of (3.26):

$$h-z\times\sqrt{\frac{h(1-h)}{n}} < p < h+z\times\sqrt{\frac{h(1-h)}{n}} \qquad (3.27)$$

The comparison of two proportions is covered in the Vaccine Example of Chapter 6. It is often more convenient in hypothesis testing to deal with the raw frequencies by means of the $\chi^2$ distribution rather than to convert them into proportions.

## Nonparametric methods

The mean and standard deviation are descriptors which apply only to metric variables. When the data is the result of a cruder level of measurement techniques relating to frequency of occurrence and rank order within the sample are employed. Sometimes the normality assumption of classical sampling theory cannot be made for one reason or another and so metric data is ranked and treated as if it were ordinal.

Analytical techniques which do not involve the estimation of parameters are called **nonparametric methods**. Such procedure are usually **distribution free** in that they do not assume a specific underlying probability distribution for the population.

As well as the sign test which follows, other nonparametric tests occur throughout this book. Many use the binomial distribution or the $\chi^2$ statistic. The main disadvantage of the distribution-free approach to statistics is that it does not provide a mathematical model of the population under study. The tests merely indicate whether a specific property is likely to be true and do not allow us to estimate terms in models which can then be tested in more detail. It is therefore difficult to employ them systematically in the step-by-step refinement process of scientific methodology.

## The sign test

The presence or absence of an attribute serves to divide a sample into two groups - items with a plus sign and items with a minus sign. For an ordinal variable, the relative sizes of the data values may have the same effect. In particular, an individual's perceptions or feelings are 'positive' or 'negative' and can be treated as an attribute variable. It is difficult to quantify a patient's perception of a treatment for the relief of pain or a viewer's reaction to a television advertisement. In such situations it is often the perception itself which is important and not the actual physiological changes that the stimulus produces.

### The hand cream example

A sample of 8 people who regularly use Verdi's Hand Cream were given a new type to try out. Five of them said that they detected an improvement in the new preparation but three said that it was the same or worse than Verdi's. Does this indicate that the new cream is significantly better?

*Solution* The data here is the presence or absence of the attribute 'improvement' for each person in the sample. The null hypothesis is that the new treatment is the same as the existing one and represents a state of disbelief which we want the data to convince us is false.

Under the null hypothesis, the probability of a reported improvement for an individual is 0.5. The question now becomes whether the occurrence of 5 'successes' in 8 repetitions is so large that the improvements are not being reported merely by chance and the new treatment does have an effect. The test is therefore one-tailed as small numbers of successes would not be significant and we would not bother to apply the test to such outcomes.

The probability of a given number of successes in a repeated sequence of identical and independent trials is found from the binomial distribution. Substituting 8 and 0.5 for $n$ and $p$ in formula (3.22):

$$P(r \text{ successes out of 8 trials}) = \frac{8!}{(8-r)!r!}(0.5)^8 \tag{3.28}$$

We now want to know which outcomes form the most extreme 5% likely group so that we can test whether '5 successes' is significant. Putting $r$ equal to 8 in (3.28) gives the result 0.0039. As this is less than 5% we continue and calculate the probability of the next most extreme outcome. Putting $r$ equal to 7 in (3.28) produces the answer 0.0313 and so the probability of the experiment yielding the result '7 or 8 improvements' simply by chance is 0.0039+0.0313 which is 0.0352. This is still less than 5% and putting $r$ equal to 6 in the equation gives 0.1094. This itself is greater than 5% and hence the significant tail of the distribution is '7 or 8 successes'. 'Any number less than 7' has a probability of about 95% and is consistent with the null hypothesis of there being no improvement. If the sample had contained 7 improvements then we would reject $H_0$ at the 3.5% level of significance because the probability of an outcome as extreme as that is 0.0352. It follows that **the 5 improvements we are told about in the question form insufficient evidence that the new cream gives a perceived benefit to patients.**

The sign test can be used to test a null hypothesis about a population median. The probability that a data value is less than the population median is 0.5 by its definition. Hence the number of times this occurs for a sample of values is the number of successes in a binomial trial with $p$ equal to 0.5. By testing whether this number is significantly large or small we can decide on the credibility of the hypothesised median. In this form the test is called the **median test** and it is also known as the **binomial test**.

For large samples the $\chi^2$ test can be applied to the frequencies of plus and minus signs in the sample. This was done in the Coin Example earlier in the chapter where 'heads' and 'tails' took the place of 'plus' and 'minus' and appears again in the McNemar test in Chapter 4. An explanation of the equivalence of the two tests follows for the reader who cares for such things.

The probability distribution of the sum of the squares of $\nu$ standard normal variables is $\chi^2$ with $\nu$ degrees of freedom. Hence when $\nu$ is equal to 1, $\chi^2$ values are simply $z^2$ values. Testing a $\chi^2$ statistic with 1 degree of freedom is therefore the same as testing its square root as a $z$ score. The test is a two-tail one as for $z^2$, which is $\chi^2$, to be large and positive, $z$ can be large and positive or large and negative. Putting all this together, we see that testing a $z$ score which, because of the normal approximation to the binomial, measures the significance of an observed number of successes, is equivalent to testing a $\chi^2$ statistic derived from that number of successes.

## Summary

**1** The normal distribution describes random variables which have symmetric bell-like histograms. It is tabulated as the area under the standard normal curve to the right of a $z$ score. Percentage points tables allow a $z$ score to be found from a knowledge of the area. Probabilities are calculated from these areas or prediction intervals can be formed. The normal distribution is additive in that the sum of any linear combination of normal variables is itself normally distributed.

**2** Classical sampling theory is concerned with random samples in which the variable of interest is normally distributed in the population. Under this normality assumption, the mean of the sample is itself normally distributed with mean equal to the population mean and standard deviation equal to the population standard deviation divided by the square root of the sample size.

**3** If the population standard deviation is unknown and must be estimated from the sample, then the sample mean is described by the $t$ distribution instead of the normal. The number of degrees of freedom for $t$ is one less than the sample size.

**4** The variance of a sample drawn from a normal population has the $\chi^2$ distribution. This has the same number of degrees of freedom as $t$. The $\chi^2$ statistic measures the extent to which a set of expected frequencies agree with a set of observed ones. The number of degrees of freedom for this purpose is the number of expected frequencies minus the number of distinct quantities calculated from the data. A correction due to Yates should be used when there is only 1 degree of freedom.

**5** Statistical inference takes the form of estimation and hypothesis testing. Samples can give point estimates or confidence interval estimates of population parameters. A confidence interval is a range within which we have a stated amount of faith that the population parameter lies.

**6** A null hypothesis is a working assertion about the value of a population parameter which we seek to disprove by means of evidence derived from the data. The alternative hypothesis is the statement we accept by rejecting the null hypothesis. A one-tail test applies when the evidence must sway our opinion in one direction away from the hypothesised value. A two-tail test causes rejection of the null hypothesis if the evidence reduces significantly the likelihood of the sample's randomness by simply being different from the hypothesised value. The level of significance of a test is a threshold probability. We do not believe that samples with probabilities lower than this have been selected by chance. The level of significance establishes acceptance and rejection regions for the test statistic so that its position indicates whether the null hypothesis should be rejected or not.

**7** The uniform probability distribution allocates equal likelihood to each outcome of a trial. The binomial distribution describes sequences of trials which are all identical and statistically independent. The Poisson distribution relates to events which occur randomly in space or time. It also acts as an approximate substitute for the binomial when the number of trials is large. The normal distribution gives approximations to them both provided the relevant parameters are large. The proportion or percentage of items in a sample or population can be described using the normal approximation to the binomial.

**8** Methods which do not involve the estimation of population parameters are called nonparametric. They are usually distribution-free in that they do not assume a specific probability distribution for the parent population as classical sampling theory does. The sign test, otherwise known as the median or binomial test, uses the binomial distribution to determine whether an attribute with two possible values takes them both equally often.

## Further reading

The analysis of a single sample by the methods of classical sampling theory occupies a key position within the art form known as practical statistics. Many applications like quality control, clinical trials and survey analysis are often implemented on single samples with attention being focussed on one variable at a time. For this reason the probability distributions introduced in this chapter recur throughout the book and should be understood before the serious reader ventures further. In view of its important role, this material on classical sampling forms a part of every one of the references in the first two sections of the Bibliography. Spiegel (10) contains many worked and unworked examples for readers wishing to gain practice in the technique concerned.

Apart from (20) and (27), which deal with the Bayesian approach, and (26) which is restricted to probability theory, all the texts in the Theoretical Statistics section give the mathematical background to classical sampling. Reference (24) is considered the authoritative work and consists of 3 volumes of very closely argued higher mathematics.

Statistics books differ in their attitude to nonparametric methods and some do not include them at all. The section on Nonparametric Methods lists specialised texts, (29) and (33) being particularly good. These techniques are incorporated in this book as and when appropriate.

## Exercises

**1** The length of an animal's body is normally distributed with mean 78 mm and standard deviation 13 mm. (i) Calculate the probability that an animal chosen at random has a body less than 84 mm long. (ii) Calculate the probability that an animal chosen at random has a body between 73 mm and 89 mm long. (iii) The animal's head is also normally distributed with mean 27 mm and standard deviation 9 mm. Use the additive property of the normal distribution to derive a 95% prediction interval for the total length of the animal. (iv) Calculate a 99% prediction interval for the average body length of a sample of 15 animals chosen at random.

**2** The time measured in man-hours taken to place a cubic metre of concrete was measured on 10 separate occasions to be 2.3, 3.1, 6.0, 4.2, 5.9, 2.8, 3.0, 4.4, 9.7 and 8.4. (i) Determine a 95% confidence interval for the population mean time to lay a cubic metre of concrete. (ii) In preparing estimates of costs in the past, a company has assumed that the average time is 6 man-hours. Does the sample indicate that the population mean is in fact different from this? (iii) Is there evidence that the population mean is bigger than 3.5 man-hours? (iv) Calculate a 95% confidence interval for the population variance of the times taken to place a cubic metre of concrete.

**3** A die was thrown 253 times giving the following outcomes:

| Face shown | 1 | 2 | 3 | 4 | 5 | 6 |
|---|---|---|---|---|---|---|
| Frequency | 31 | 52 | 48 | 39 | 50 | 33 |

Perform a statistical test to decide whether the die is fair.

**4** The Hardy–Weinberg Law of genetics states that the probabilities of different gene combinations are given by the binomial distribution. Consider the data from a sample of 147 flowers:

| Genotype | XX | XY or YX | YY |
|---|---|---|---|
| Frequency | 24 | 68 | 55 |

(i) Add up the total number of X genes in the sample. (ii) Estimate the probability of a gene chosen at random being an X gene by dividing this number by the total number of genes, 294. (iii) Calculate the probabilities of 0, 1 and 2 successes out of 2 trials with the probability of success given by the answer to (ii) using the binomial distribution. (iv) Multiply each of the three probabilities obtained above by 147 to derive expected frequencies to compare with the ones observed. (v) Perform a goodness-of-fit test to decide whether the Hardy–Weinberg Law describes the observations.

**5** A furniture store experiences the following demand for beds:

| Number of beds requested per week | 0 | 1 | 2 | 3 | 4 or more |
|---|---|---|---|---|---|
| Number of weeks | 22 | 47 | 35 | 18 | 11 |

If demand occurs randomly in time then the Poisson distribution should describe the observations. Test this hypothesis.

**6** A traffic survey shows that on a particular day 328 of the 904 vehicles passing through Ditchpuddle were heavy lorries. (i) Calculate a 90% confidence interval for the population proportion. (ii) Is the population proportion no bigger than 30%?

**7** In a controlled experiment a woman guessed the suit of a playing card chosen at random in another room on 29 occasions out of 85 trials. (i) If the woman is guessing randomly then the probability of success for each trial is 0.25. Use the normal approximation to the binomial distribution to determine a 95% prediction interval for the number of correct guesses out of 85 trials based on this null hypothesis. (ii) The woman has attained a level of success inside the range of your prediction interval. However, this is a two-tail approach and we want to know if her success rate is significantly better than merely guessing. Perform a suitable test either on the number of successes or on the proportion of successes to examine this null hypothesis. (iii) Describe how the two hypotheses mentioned above can be tested by the $\chi^2$ goodness-of-fit test.

**8** An economist interviews 14 employees doing similar jobs and finds that 5 of them have salaries below £280 per week with the remainder being above. Perform a sign test on the null hypothesis that the median of the population is £280.

# 4 One sample with two variables

When two different quantities are measured on each item in a sample we are usually interested in the degree of association or correlation between them. If this is not the case then the techniques of the previous chapter can be used to analyse each variable separately. The first two examples in this chapter utilise the $\chi^2$ distribution introduced in Chapter 3 to determine whether there is an association between two attribute variables. The subject of the third and fourth examples is regression analysis. This is an attempt to find a mathematical relationship between two metric variables. The last section of the chapter deals with situations where the two variables are paired in some way.

## Contingency tables

### The cars example

A garage repairs 600 cars during July, August and September. The manager records the manufacturer and type of fault for each one:

| | Manufacturer | | | |
|---|---|---|---|---|
| Type of fault | Allgood | Breakwell | Cruiser | Drivealong |
| Electrical | 69 | 57 | 70 | 61 |
| Mechanical | 50 | 61 | 56 | 49 |
| Bodywork | 15 | 58 | 20 | 34 |

*Question 1* Is there an association between the make of car and the type of fault it develops?

*Solution* The table is a 3 rows by 4 columns **contingency table.** The 600 cars have been **cross-classified** or **cross-tabulated** according to the two attributes 'make' and 'type of fault'. The question is whether one method of classification is related to or associated with the other. For instance there might be a tendency for certain makes of vehicle to a certain type of fault. The null hypothesis we assume is that there is no association between the variables at all.

We begin a test of this null hypothesis by calculating the frequencies we would have expected to observe if it were true. Table 4.1 contains the frequencies from the question with the rows and columns totalled. The expected frequencies are written in brackets underneath the observed ones.

As an example of the calculation of the expected frequencies consider that for Cruiser cars having bodywork faults. From the column total we see that the fraction (146/600) of the cars were made by Cruiser. The null hypothesis implies that this same fraction of the 127 cars with bodywork faults should have been made by Cruiser. In other words the proportion of Cruiser cars with bodywork faults should be the same as the proportion of Cruiser cars in the population as a whole. Hence the expected frequency for Cruiser cars with bodywork faults is $(146 \times 127/600)$, which is 30.9. The other expected frequencies are calculated in the same way, by simply multiplying the appropriate row and column totals together and dividing by the grand total, 600. Notice that these frequencies have the same row and column totals as the original observations. This is a useful check on the accuracy of the calculations.

**Table 4.1** Contingency table for the Cars Example

| | Manufacturer | | | | |
|---|---|---|---|---|---|
| Type of fault | Allgood | Breakwell | Cruiser | Drivealong | Total |
| Electrical | 69<br>(57.4) | 57<br>(75.4) | 70<br>(62.5) | 61<br>(61.7) | 257 |
| Mechanical | 50<br>(48.2) | 61<br>(63.4) | 56<br>(52.6) | 49<br>(51.8) | 216 |
| Bodywork | 15<br>(28.4) | 58<br>(37.3) | 20<br>(30.9) | 34<br>(30.5) | 127 |
| Total | 134 | 176 | 146 | 144 | 600 |

Having determined the expected frequencies as consequences of the null hypothesis, we now decide whether the observations are sufficiently close to them to be considered consistent with that hypothesis. In the Newspaper Example and others of Chapter 3 two sets of frequencies were tested for similarity with the $\chi^2$ goodness of fit test. None of the expected frequencies should be less than 5 for such a test and in our contingency table this is the case. The procedure for merging categories when the frequencies are small will be covered in the next example.

Applying equation (3.17) to our data:

$$\chi^2 = \frac{(69-57.4)^2}{57.4} + \frac{(57-75.4)^2}{75.4} + \frac{(70-62.5)^2}{62.5} + \frac{(61-61.7)^2}{61.7} + \frac{(50-48.2)^2}{48.2} + \cdots + \frac{(34-30.5)^2}{30.5} = 30.33 \tag{4.1}$$

A relatively large value of $\chi^2$ would indicate that the observed and expected frequencies are dissimilar. This would mean that the null hypothesis does not predict the observations and there is an association of some sort between the variables.

The size of $\chi^2$ is interpreted with respect to its degrees of freedom. According to equation (3.19) we need the number of expected frequencies minus the number of quantities calculated from the observations. In a contingency table the rows and columns are totalled. This means that any individual row or column can be reconstructed from a knowledge of the totals and the remaining frequencies. Effectively the table has been reduced in information content by one row and one column:

$$\text{Degrees of freedom of a contingency table} = (\text{number of rows} - 1) \times (\text{number of columns} - 1) \qquad (4.2)$$

If there is only one degree of freedom then Yates' correction should be used as described in Chapter 3, but in our example the product is $(3-1)\times(4-1)$ which is 6. We find from the tables on page 188 that the 5% significant value of $\chi^2$ for 6 degrees of freedom is 12.5916 and so our answer of 30.33 is relatively very large. Looking along the row of the table corresponding to 6 degrees of freedom, we see that it is significant at the 1% level. We therefore conclude that **there is an extremely strong association between the make of a car and the type of fault it develops.**

In reducing all the data to a $\chi^2$ statistic we are clearly in danger of missing any subtle messages the figures may have for us. Having established the existence of an association we can now attempt to explain whereabouts in the table it occurs. It appears that the biggest discrepancies between observation and expectation are in the first three columns of the first and last rows. It seems fair to say that **Breakwell cars have more bodywork faults than is to be expected by chance while Allgood and Cruiser cars have more electrical faults. The remainder of the frequencies could well have arisen by chance.**

*Question 2* Is there evidence that the makes of cars are significantly different from each other in the overall number of faults they develop?

*Solution* The row and column totals of a contingency table describe the occurrence of the two variables independently of each other. In this example, the column totals are the number of cars classified by manufacturer only. They tell us how the faults were distributed according to the single attribute 'manufacturer'. This single sample with one variable data structure is the same as that for the Newspaper Example in Chapter 3. The question suggests that we test the null hypothesis that all makes of car are equally prone to develop a fault and hence the number of faults should all be the same.

As there are 600 cars in the sample, the hypothesis implies that each of the four manufacturers should have been responsible for 150 faults. The $\chi^2$ statistic is:

$$\chi^2 = \frac{(134-150)^2}{150} + \frac{(176-150)^2}{150} + \frac{(146-150)^2}{150} + \frac{(144-150)^2}{150} = 6.56 \qquad (4.3)$$

The appropriate number of degrees of freedom for this statistic is 3 because there are 4 frequencies predicted and one calculation performed, their addition together. The table on page 188 gives the 5% significant value of $\chi^2$ with 3 degrees of freedom to be 7.8147. The computed value is smaller than this and so **there is no evidence of a significant difference between the various makes of cars in the number of faults they develop.**

Contingency tables can relate ordinal and metric variables to each other as well as to attribute ones. In the next example the classes of a metric variable are treated as attributes for this purpose. It also illustrates how expected frequencies which are less than 5 are dealt with.

### The drinks example

A survey on the type of non-alcoholic drink preferred by 288 males gave the following results:

| | Age in years | | | | | Total |
|---|---|---|---|---|---|---|
| | 0–5 | $5^+$–10 | $10^+$–20 | $20^+$–50 | $50^+$ | |
| Milk | 8 | 4 | 12 | 0 | 0 | 24 |
| Lemon | 1 | 2 | 1 | 4 | 8 | 16 |
| Orange | 6 | 16 | 13 | 10 | 8 | 53 |
| Cola | 1 | 9 | 15 | 12 | 3 | 40 |
| Tea | 2 | 7 | 16 | 21 | 18 | 64 |
| Coffee | 0 | 6 | 19 | 17 | 16 | 58 |
| Other | 0 | 7 | 5 | 12 | 9 | 33 |
| Total | 18 | 51 | 81 | 76 | 62 | 288 |

This type of contingency table arises very often in the analysis of questionnaire data. It is the result of cross-tabulating the responses to two questions, one on age and the other on preferred drink. It tells us, for instance, that 7 respondents indicated 'tea' as their favourite drink and 'over 5 but not over 10' as their age.

*Question 1* Is there an association between age and type of drink preferred?

*Solution* Before calculating the $\chi^2$ statistic we check that none of the expected frequencies are less than 5. The smallest of them will correspond to the smallest totals because they are the products of the row and column totals divided by the grand total. In this case it is $(18 \times 16/288)$, which is 1 and unacceptably low. However, if we can merge either the row or the column we have just identified with another one, the total will increase and the expected frequency will be higher. Naturally any regrouping of the data in this way must be justifiable within the context of the situation being investigated. For instance combining 'lemon' with 'milk' would eliminate the smallest row total but create the category of favourite drink 'milk or lemon'. This does not seem a meaningful description of a type of drink. It is more sensible to merge columns as this merely reclassifies the age variable into bigger classes, but in this case the low row total for 'lemon' would still need attention even if the first two columns were merged. Similarly, merging 'lemon' with the category 'other' does not give all the expected frequencies to be at least 5. Perhaps it should be pointed out at this stage that some statisticians would continue the analysis in spite of the low expected frequencies. Provided they do not affect a sizeable part of the table they are not considered to be too much of a hindrance. As always in a statistical analysis, nothing is final and we could try the analysis in different ways.

Let us suppose that the researcher is prepared to consider 'lemon' and 'orange' as a new category, 'fruit'. The reduced table has smallest column total equal to 18 but now the total for 'milk', 24, is the smallest row one. To avoid merging 'milk' with another row we reclassify the age variable and merge the first two columns together. The reduced contingency table is shown as Table 4.2.

The $\chi^2$ statistic for this table is

$$\chi^2 = \frac{(12-5.75)^2}{5.75} + \frac{(12-6.75)^2}{6.75} + \frac{(0-6.33)^2}{6.33} + \cdots + \frac{(9-7.10)^2}{7.10} = 49.47 \qquad (4.4)$$

The number of degrees of freedom according to equation (4.2) is $(6-1) \times (4-1)$, which is 15. From the tables we see that the 5% critical value for $\chi^2$ with this number of

**Table 4.2** Reduced contingency table for the Drinks Example

| | Age in years | | | | |
|---|---|---|---|---|---|
| | 0–10 | $10^+$–20 | $20^+$–50 | $50^+$ | Total |
| Milk | 12 (5.75) | 12 (6.75) | 0 (6.33) | 0 (5.17) | 24 |
| Fruit | 25 (16.53) | 14 (19.41) | 14 (18.21) | 16 (14.85) | 69 |
| Cola | 10 (9.58) | 15 (11.25) | 12 (10.56) | 3 (8.61) | 40 |
| Tea | 9 (15.33) | 16 (18.00) | 21 (16.89) | 18 (13.78) | 64 |
| Coffee | 6 (13.90) | 19 (16.31) | 17 (15.31) | 16 (12.49) | 58 |
| Other | 7 (7.91) | 5 (9.28) | 12 (8.71) | 9 (7.10) | 33 |
| Total | 69 | 81 | 76 | 62 | 288 |

degrees of freedom is 24.9958. Following the row along in the table, it seems that our calculated value is very highly significant. **There is strong evidence of an association between age and type of non-alcoholic drink preferred by males.**

Looking at the relative sizes of the observed and expected frequencies in order to try and explain this result, we see that the largest discrepancies are for milk drinkers. There are too many of them under 20 and too few over 20 to have occurred by chance alone, and this suggests that the table be re-analysed ignoring the milk row altogether.

A contingency table can be **partitioned** and rows and columns can be amalgamated for certain tests. The researcher should beware of being led by the nature of the particular sample selected in this 'follow-up' analysis. It is easy to concoct null hypotheses which are in complete accord or discord with the data values, and so engage in *post hoc* or 'after the event' reasoning. The spirit of statistical method is that

**Table 4.3** Contingency table for Question 2 of the Drinks Example

| | Type of drink | | | | | | |
|---|---|---|---|---|---|---|---|
| Age | Milk | Fruit | Cola | Tea | Coffee | Other | Total |
| 0–10 | 12 (5.75) | 25 (16.53) | 10 (9.58) | 9 (15.33) | 6 (13.90) | 7 (7.91) | 69 |
| $10^+$ | 12 (18.25) | 44 (52.47) | 30 (30.42) | 55 (48.67) | 52 (44.10) | 26 (25.09) | 219 |
| Total | 24 | 69 | 40 | 64 | 58 | 33 | 288 |

hypotheses are formulated before any data is collected and we are guided one way or another in our conclusions about those hypotheses by the content of the sample.

*Question 2* Is there evidence that respondents in the 'under 10' age group have a significantly different set of preferences from the rest of the sample?

*Solution* Table 4.3 can be obtained by merging all the age categories except the first. The $\chi^2$ statistic for this table is 24.15 and it has $(2-1)\times(6-1)$ degrees of freedom. The 5% significant value of $\chi^2$ corresponding to 5 degrees of freedom is 11.0705 and so **there is very strong evidence of an association between the type of drink preferred and whether the respondent was over or under the age of 10.**

## Correlation and regression

The contingency tables examined in the last two examples summarised the correlation between two attribute or classified metric variables. When both variables have metric scales the possibility of a mathematical relationship between them can be investigated.

### The cholesterol example

The concentrations of cholesterol in the blood serum of 12 women were recorded together with their ages:

| Case | 1 | 2 | 3 | 4 | 5 | 6 | 7 | 8 | 9 | 10 | 11 | 12 |
|---|---|---|---|---|---|---|---|---|---|---|---|---|
| Age (years) | 42 | 49 | 64 | 33 | 71 | 55 | 49 | 52 | 37 | 59 | 50 | 62 |
| Cholesterol (mg/100 ml) | 151 | 170 | 226 | 147 | 254 | 235 | 194 | 165 | 153 | 218 | 222 | 246 |

*Question 1* Is there a mathematical relationship between the concentration of cholesterol in the blood and age?

*Solution* In the Gulls Example of Chapter 2 the relationship between two metric variables was illustrated by a **scatter diagram.** Each item in the sample is represented by a point whose coordinates are the variables being measured. There are three types of patterns that the points on a scatter diagram can have. Figure 2.15 for the gulls shows **zero correlation,** for which the points form a shapeless cloud and there is no connection between the variables. The present example, depicted in Fig. 4.1, suggests **positive correlation,** in which an increase in either of the two variables implies an increase in the other one. Finally, Fig. 4.2 is an example of variables showing **negative correlation,** in which an increase in either of the two variables implies a decrease in the other one. Notice that the points do not have to lie on a straight line for correlation to be present. The terms positive and negative correlation merely describe the general behaviour of each variable when the other one changes. They do not indicate that a specific type of curve can be drawn through the points.

When a scatter diagram is drawn as a prelude to fitting a model in which one of the variables is to be predicted from values of the other, care should be taken over the choice of axes. In viewing a graph it is more natural to ascribe error to the vertical positioning of a point than to its horizontal positioning. Hence if it is felt that there is more statistical or experimental error in one of the variables than the other,

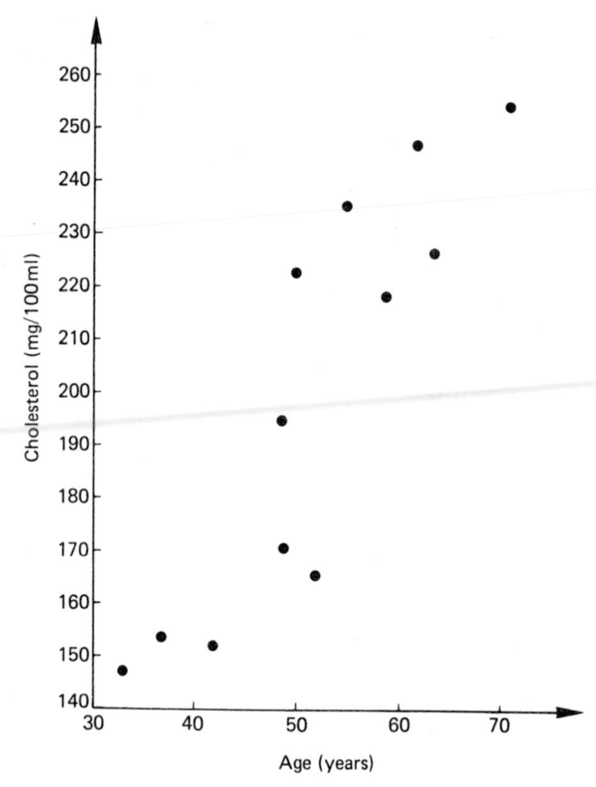


**Fig. 4.1** Scatter Diagram for the Cholesterol Example

then that one should be plotted as the vertical axis in the diagram. Similarly if one variable is known to act as the cause of the other then it should be plotted horizontally. This is because the visual impression created by a graph is that the quantity plotted vertically depends in some way on the value of the horizontal coordinate.

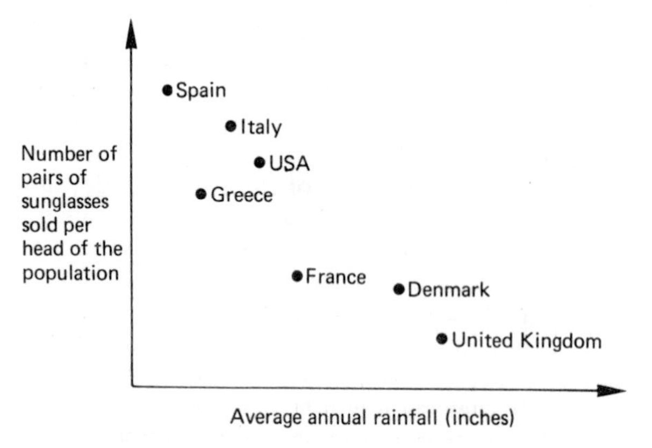


**Fig. 4.2** Example of Scatter Diagram showing negative correlation

In the Cholesterol Example there is clearly more error involved in measuring cholesterol level than age. Furthermore it seems more sensible to model the concentration as being dependent on age than to treat them the other way around. We therefore conclude from Fig. 4.1 **that it seems likely there is a mathematical relationship between the cholesterol concentration in the blood and age.**

Having drawn a scatter diagram showing non-zero correlation it is tempting to infer that a degree of causality exists between the variables. Perhaps an example will expose the danger of jumping to such a conclusion.

The number of divorces registered each year in a country fluctuates according to the size of the country's population. Similarly the annual sales of ice cream is proportional to the population size. Consequently a scatter diagram of the number of divorces in a year against the sales of ice cream for that year would show a positive correlation between them. If this is interpreted as evidence of 'cause and effect', then either eating ice cream causes marital disharmony or getting divorced makes people eat ice cream! In fact the appearance of the scatter diagram and the correlation it shows are completely explained by the existence of the 'hidden' third variable, population size. Whenever two variables are strongly related to a third variable then they themselves will exhibit a **spurious correlation** with each other.

It follows that the validity of a conclusion regarding causality cannot be established by a statistical argument alone. Whilst it is a necessary prerequisite for causality that the correlation between the variables is non-zero, the branch of science appropriate to the problem in hand must provide its own supporting justification for the purely mathematical treatment of the data.

There are situations in which spurious correlation is turned to advantage. In forecasting applications in operational research we may be able to forecast values for one variable and, because of their correlation, produce forecasts for a second variable. It does not matter whether the quantities have a causal relationship, only that there is a mathematical equation to generate values of one from values of the other.

*Question 2* Determine the precise form of the relationship.

*Solution* This is an obvious extension to the previous question. Having decided that the points on a scatter diagram lie on some sort of curve it is reasonable to want to know its equation. This will in fact constitute a model of the $y$ variable in terms of the $x$ variable.

The search for and testing of this model is called a **regression analysis.** The connection assumed is that $y$ is some function of $x$ plus or minus statistical error. All the experimental error is thus assumed to be in the $y$ measurements with the $x$ values being exact. The regression model implies that $x$, the **independent variable**, causes or predicts $y$, the **dependent variable**. As the points in Fig. 4.1 appear to lie in approximately in a straight line we shall perform a **linear regression**. This is an attempt to fit the model:

$$y = \alpha + \beta x + \text{error} \tag{4.5}$$

where the error term is normally distributed with mean equal to zero. Equation 4.5 describes the regression of **$y$ on $x$.** The fitting of curves other than straight lines, called **non-linear** or **curvilinear regression**, will be dealt with in the next example.

For theoretical reasons the best technique for estimating the parameters $\alpha$ and $\beta$ in the model is the **method of least squares**. Their values are calculated so that the sum of all the squared errors for the sample when the model is used on the data is as small as possible. The working for this example and the general formulae are in Table 4.3. **Corrected sums** were introduced in Chapter 2 for calculating variances which are

**Table 4.3** Regression line calculation for the Cholesterol Example

| | $x$ | $y$ | $x^2$ | $y^2$ | $xy$ |
|---|---|---|---|---|---|
| | 42 | 151 | 1 764 | 22 801 | 6 342 |
| | 49 | 170 | 2 401 | 28 900 | 8 330 |
| | 64 | 226 | 4 096 | 51 076 | 14 464 |
| | 33 | 147 | 1 089 | 21 609 | 4 851 |
| | 71 | 254 | 5 041 | 64 516 | 18 034 |
| | 55 | 235 | 3 025 | 55 225 | 12 925 |
| | 49 | 194 | 2 401 | 37 636 | 9 506 |
| | 52 | 165 | 2 704 | 27 225 | 8 580 |
| | 37 | 153 | 1 369 | 23 409 | 5 661 |
| | 59 | 218 | 3 481 | 47 524 | 12 862 |
| | 50 | 222 | 2 500 | 49 284 | 11 100 |
| | 62 | 246 | 3 844 | 60 516 | 15 252 |
| *Total* | 623 | 2 381 | 33 715 | 489 721 | 127 907 |

Sample size, $n=12$

$$\bar{x}=\frac{\text{sum of } x}{n}=\frac{623}{12}=51.9167$$

$$\bar{y}=\frac{\text{sum of } y}{n}=\frac{2\,381}{12}=198.4167$$

Corrected sum of $x^2$, $CS_{xx}=\text{sum of }(x^2)-((\text{sum of } x)^2/n)$

$$=33\,715-(623^2/12)=1370.9167$$

Corrected sum of $y^2$, $CS_{yy}=\text{sum of }(y^2)-((\text{sum of } y^2)/n)$

$$=489\,721-(2381^2/12)=17\,290.9167$$

Corrected sum of $xy$, $CS_{xy}=\text{sum of }(xy)-((\text{sum of } x)(\text{sum of } y)/n)$

$$=127\,907-(623\times 2381/12)=4293.4167$$

Estimate of $\beta$ = **Slope of regression line,** $\boldsymbol{b}=\dfrac{\boldsymbol{CS_{xy}}}{\boldsymbol{CS_{xx}}}=\dfrac{4293.4167}{1370.9167}=3.1318$

Estimate of $\alpha$ = **Intercept of regression line,** $\boldsymbol{a=\bar{y}-b\bar{x}}$

$=198.4167-3.1318\times 51.9167=35.8240$

**Regression line is $\boldsymbol{y=35.8+3.13x}$**

corrected sums of squares divided by the sample size minus one. The data values can be scaled up or down before the calculations are performed in order to keep these sums to a manageable size. This is illustrated in the next example. Figure 4.3 shows the scatter diagram with the regression line plotted. It is necessary to plot only two points in order to draw a straight line. From the calculated equation we see that for $x$ equal to 40, say, $y$ is equal to 161.0. Plotting this point and joining it to the point with coordinates $\bar{x}$ and $\bar{y}$, which always lies on the line, produces the graph.

Sometimes the points on a scatter diagram are in such an obvious straight line that we can use a ruler to draw the line without performing any calculations. The fact that $(\bar{x}, \bar{y})$ is on the 'best' line can be utilised in positioning the ruler when this procedure

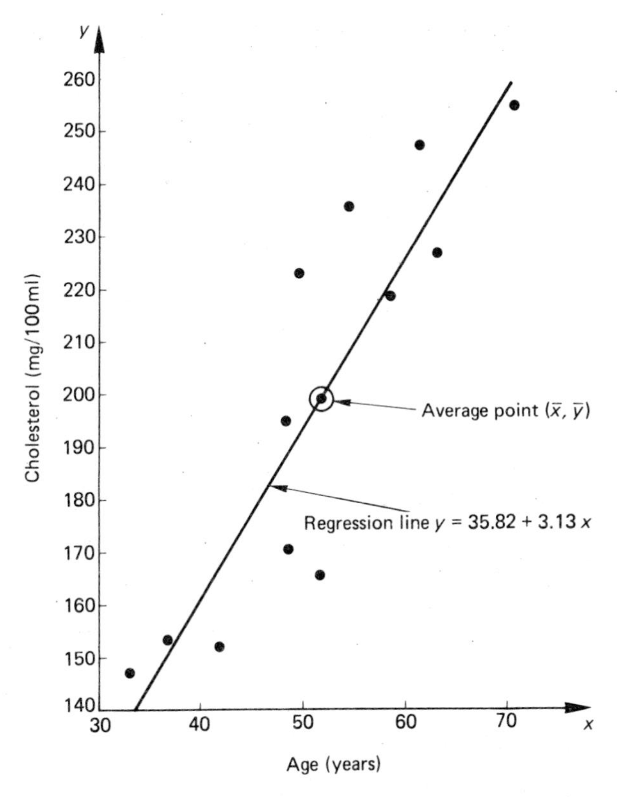


**Fig. 4.3** Regression line for the Cholesterol Example

is adopted. Note that when measuring lengths to estimate the gradient of a line drawn by inspection, the scales of the axes should be used and not the readings from a ruler. The units of the slope of a regression line are 'units of $y$ per unit of $x$' and in our example this is 'mg per 100 ml per year of age'. The intercept, being a measurement on the $y$ axis, has the same units as the dependent variable.

*Question 3* Does the equation found above give a statistically acceptable description of the sample?

*Solution* The number-crunching contained in Table 4.3 can be performed on any set of data values whether the scatter diagram appears to justify a straight-line model or not. It is therefore necessary to test the validity of the model as a descriptor of the relationship between the variables.

A popular way of doing this is to calculate the correlation coefficient:

$$r = \frac{\text{covariance of } x \text{ and } y}{(\text{standard deviation of } x)(\text{standard deviation of } y)} = \frac{CS_{xy}}{\sqrt{(CS_{xx}CS_{yy})}} \qquad (4.6)$$

where the covariance of $x$ and $y$ is the corrected sum of $xy$ divided by $(n-1)$. It is an extension of the concept of variance to measure the combined variability of two variables considered together. An alternative formula for the slope of the regression line is the covariance divided by the variance of $x$. However, the versions in Table 4.3

reduce the possibility of numerical error as they refer directly to the corrected sums. We shall use them for calculations rather than intermediate quantities like standard deviations. Formulae like (4.6) will be given in a form the reader may encounter in other textbooks and then expressed in terms of the corrected sums after a second 'equals' sign.

In spite of its name, the correlation coefficient measures the degree of linear dependence and not the level of general correlation. It has the value +1 if $x$ and $y$ are perfectly positively correlated with all the points on the scatter diagram lying on a straight line. It is small if $x$ and $y$ are uncorrelated while it is $-1$ if they are perfectly negatively correlated with a linear relationship. The coefficient is always between the extremes of $-1$ and $+1$ and is interpreted according to its whereabouts on that scale.

The correlation coefficient for our example is

$$r = \frac{4293.4167}{\sqrt{(1370.9167 \times 17\,290.9167)}} = +0.8818 \qquad (4.7)$$

**This result indicates a high degree of positive correlation between cholesterol level and age and justifies the fitting of a straight-line model to the data.**

The correlation coefficient (4.6) and the slope of the fitted regression line in Table 4.3 both depend on the corrected sum of $xy$, $CS_{xy}$. The null hypothesis that the correlation coefficient is zero and the variables are uncorrelated is thus logically equivalent to the hypothesis that the slope is zero. Now the sample estimate $b$ has the $t$ distribution with mean equal to the population parameter $\beta$. The statistic

$$t = (b - \beta)\sqrt{\frac{(n-2)CS_{xx}}{(1-r^2)CS_{yy}}} = (CS_{xy} - \beta CS_{xx})\sqrt{\frac{n-2}{CS_{xx}CS_{yy} - CS_{xy}^2}} \qquad (4.8)$$

has the $t$ distribution with $(n-2)$ degrees of freedom and allows us to test whether $b$ is a random fluctuation from some hypothesised value of $\beta$ or not. Usually we test for $\beta$ being equal to zero as this is equivalent to assessing the correlation coefficient. The expression for $t$ in this case reduces to

$$t = r\sqrt{\frac{n-2}{1-r^2}} = CS_{xy}\sqrt{\frac{n-2}{CS_{xx}CS_{yy} - CS_{xy}^2}} \qquad (4.9)$$

The reason the $t$ distribution appears on the scene is that in assuming the error term in the model (4.5) is normally distributed we are invoking classical sampling theory as described in the last chapter.

Testing the null hypothesis that $\beta$ is zero for our sample using (4.9):

$$t = 4293.4167\sqrt{\frac{12-2}{(1370.9167)(17\,290.9167) - (4293.4167)^2}} = 5.914 \qquad (4.10)$$

The 5% critical $t$ value for 10 degrees of freedom is 2.228 in a two-tail test, from page 192. This provides the formal support for our earlier conclusion that **the variables are highly correlated and the straight-line model is extremely appropriate.**

If we had been looking for evidence of specifically positive or specifically negative correlation, then a one-tail test would have been undertaken. We used a two-tail test to establish a significant deviation from zero correlation in either direction, positive or negative.

Formula (4.8) in one of the two forms quoted can be transposed to make $\beta$ the subject. By similar mathematical trickery to that employed in the Rice and Train Journey Examples of the last chapter, a confidence interval for $\beta$ can be constructed from the corresponding one for $t$. The final expression is

$$\frac{CS_{xy}}{CS_{xx}}-\frac{t}{CS_{xx}}\sqrt{\frac{CS_{xx}CS_{yy}-CS_{xy}^2}{n-2}}<\beta<\frac{CS_{xy}}{CS_{xx}}+\frac{t}{CS_{xx}}\sqrt{\frac{CS_{xx}CS_{yy}-CS_{xy}^2}{n-2}} \quad (4.11)$$

The conscientious reader may care to verify from this pair of inequalities that the 95% confidence interval for the slope of the regression line is from 1.95 to 4.31. As this does not include the null hypothesis value zero we can draw the same conclusion as before that the variables are highly linearly correlated. The $t$ statistic (4.8) or equivalently the confidence interval (4.11) allows us to test any specified slope value for credibility. This situation arises when the regression line is supposed to verify a relationship between the variables which we had prior knowledge of, for instance a law of physics or chemistry.

*Question 4* Estimate the cholesterol level of a 55-year-old woman.

*Solution* We can obtain a point estimate by simply substituting 55 for $x$ in the regression equation. The estimate is 35.8+(3.13)(55) which is 208.0 mg/100 ml. The formula for a prediction interval of a single $y$ value is

$$a+bx\pm t\sqrt{\left[1+\frac{1}{n}+\frac{(x-\bar{x})^2}{CS_{xx}}\right)\left(\frac{CS_{yy}-CS_{xy}^2/CS_{xx}}{n-2}\right)\Bigg]} \quad (4.12)$$

where the $t$ value has $(n-2)$ degrees of freedom. The 95% prediction interval for $y$ when $x$ is equal to 55 is

$$35.8+(3.13)(55)\pm 2.228\sqrt{\left[\left(1+\frac{1}{12}+(55-51.9167)^2/1370.9167\right)\times\left(\frac{17\,290.9167-(4293.4167)^2/1370.9167}{12-2}\right)\right]}$$

$$=208.0\pm 45.6 \quad (4.13)$$

Hence **the point estimate is 208 mg/100 ml and we are 95% certain it will be between 162 mg/100 ml and 254 mg/100 ml**. A prediction interval can be used to indicate whether a particular $y$ value obeys the regression model or is an 'outlier'.

We have performed an **interpolation** by making a prediction based on an $x$ value within the range of the sample. Care should be taken when using $x$ values outside this range to carry out an **extrapolation**. The validity of a prediction depends on the underlying assumptions of the model being true and this may not be so beyond the range of the data.

*Question 5* Derive a confidence interval for the mean of the cholesterol levels of all women aged 55 years.

*Solution* This question is not about the original population of women of all ages but about the population of women aged 55. A point estimate for the mean cholesterol level within this population is found from the regression line and is the same as the answer to Question 4. A confidence interval for it is given by

$$a+bx\pm t\sqrt{\left[\left(\frac{1}{n}+\frac{(x-\bar{x})^2}{CS_{xx}}\right)\left(\frac{CS_{yy}-CS_{xy}^2/CS_{xx}}{n-2}\right)\right]} \quad (4.14)$$

The formula is similar to (4.12) except that the figure '1' is absent from the square root. The $t$ value again has $(n-2)$ degrees of freedom. Putting our sample values in the equation gives the interval **195 mg/100 ml to 221 mg/100 ml**. Note that this interval is smaller than the prediction interval calculated in the previous solution. We are more confident about the whereabouts of the mean than the whereabouts of one particular $y$ value.

## Curvilinear regression

The relationship $y = \alpha + \beta x$ is a linear one as it produces a straight-line graph. In many practical problems the points on a scatter diagram do not lie on or near to a straight line. In some cases we want to discover the type of relationship which exists and in others we suspect a specific equation to be true. The general method is to transform the original variables to new ones using suitable functions like logarithms, squares and square roots. The connection between the new variables is then assumed to be linear and the techniques described above which are readily available on computer packages and have an extensive theoretical underpinning can be employed. Regressions can be performed very quickly on a computer. A trial and error approach can be adopted until suitable transformations are found. Generating extra variables which are functions of the independent variable adds more terms to the model and necessitates a multivariate regression, which is covered in Chapter 5. This chapter continues with an example of a non-linear model with just one independent variable.

### The heater example

The temperatures near an electric heater as different amounts of current are fed through it have been measured:

| Current ($I$ amperes) | 1.0 | 1.2 | 1.4 | 1.6 | 1.8 | 2.0 | 2.2 | 2.4 | 2.6 |
|---|---|---|---|---|---|---|---|---|---|
| Temperature ($T$°C) | 124.1 | 112.5 | 135.4 | 121.7 | 142.4 | 126.9 | 147.1 | 156.3 | 152.7 |

There are reasons to suspect that the relationship $T = AI^2 + B$ holds where $A$ and $B$ are constants.

*Question 1* Plot a suitable scatter diagram to decide whether this type of equation could describe the data values.

*Solution* We first identify the current as the controlled independent variable and the temperature as the resulting dependent variable. If the suspected model is correct, then plotting $T$ values against $I$ values will not produce points in an approximate straight line. The equation represents a curve whose exact shape would be impossible to draw without knowledge of $A$ and $B$. When the relationship is linear we can move a ruler around on the scatter diagram in order to fit the most appropriate line or we can inspect the diagram to assess the suitability of the straight-line model for the data. There is no corresponding method of inspecting the range of possible curves in the non-linear case.

The new variables $x = I^2$ and $y = T - 100$ have the effect of **linearising** the suspected relationship because it now means that $y = \alpha + \beta x$ with $\alpha$ equal to $(B - 100)$ and $\beta$ equal to $A$. The subtraction of 100 from each temperature to form the $y$ value will help to keep the size of the numbers in the subsequent calculations within reasonable bounds. It is often possible to scale variables like this to reduce the corrected sums which have to be handled. The original variables are substituted back into the $x$ and $y$ model at the end of the analysis.

The values of the transformed variables can be seen in Table 4.4 and a scatter diagram in Fig. 4.4. **The diagram indicates that a straight-line model would be appropriate for the transformed variables.**

*Question 2* Estimate the values of the parameters $A$ and $B$ by performing a linear regression on the new variables.

*Solution* Table 4.4 contains the analysis and it is directly comparable with Table 4.3 for the Cholesterol Example. When the original variables are substituted back in the regression equation from the table:

$$T = 113.09 + 6.38I^2 \tag{4.15}$$

**Our estimates are that A is 6.38 and B is 113.09.**

**Table 4.4** Regression calculation for the Heater Example

| $I$ | $T$ | $x$ ($I^2$) | $y$ ($T-100$) | $x^2$ | $y^2$ | $xy$ |
|---|---|---|---|---|---|---|
| 1.0 | 124.1 | 1.0 | 24.1 | 1.0000 | 580.81 | 24.100 |
| 1.2 | 112.5 | 1.44 | 12.5 | 2.0736 | 156.25 | 18.000 |
| 1.4 | 135.4 | 1.96 | 35.4 | 3.8416 | 1 253.16 | 69.384 |
| 1.6 | 121.7 | 2.56 | 21.7 | 6.5536 | 470.89 | 55.552 |
| 1.8 | 142.4 | 3.24 | 42.4 | 10.4976 | 1 797.76 | 137.376 |
| 2.0 | 126.9 | 4.00 | 26.9 | 16.0000 | 723.61 | 107.600 |
| 2.2 | 147.1 | 4.84 | 47.1 | 23.4256 | 2 218.41 | 227.964 |
| 2.4 | 156.3 | 5.76 | 56.3 | 33.1776 | 3 169.69 | 324.288 |
| 2.6 | 152.7 | 6.76 | 52.7 | 45.6976 | 2 777.29 | 356.252 |
| Total | | 31.56 | 319.1 | 142.2672 | 13 147.87 | 1 320.516 |

Sample size, $n=9$; $\bar{x} = \dfrac{\text{sum of } x}{n} = \dfrac{31.56}{9} = 3.5067$

$\bar{y} = \dfrac{\text{sum of } y}{n} = \dfrac{319.1}{9} = 35.4556$

Corrected sum of $x^2 = CS_{xx} = 142.2672 - \dfrac{(31.56)^2}{9} = 31.5968$

Corrected sum of $y^2 = CS_{yy} = 13\,147.87 - \dfrac{(319.1)^2}{9} = 1834.0022$

Corrected sum of $xy = CS_{xy} = 1320.516 - \dfrac{(31.56)(319.1)}{9} = 201.5387$

Slope of regression line, $b = \dfrac{CS_{xy}}{CS_{xx}} = \dfrac{201.5387}{31.5968} = 6.3785$

Intercept, $a = \bar{y} - b\bar{x} = 35.4556 - (6.3785)(3.5067) = 13.0881$

Regression line is $y = 13.09 + 6.38x$

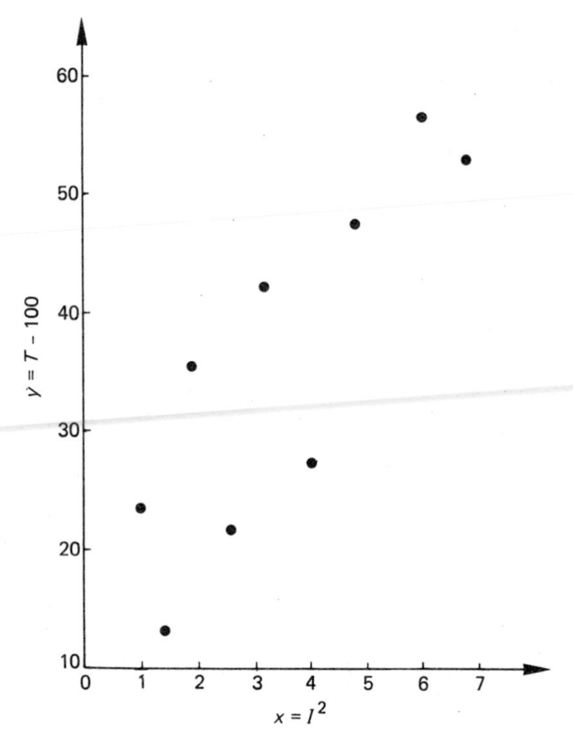


**Fig. 4.4** Scatter Diagram of the transformed variables in the Heater Example

*Question 3* Test the significance of the regression.

*Solution* In the last example the hypothesis that the slope is equal to zero was examined by means of a $t$ statistic. This is equivalent to testing the correlation coefficient for significance and could be used here. An alternative but mathematically identical method is to perform an **analysis of variance** or **ANOVA**.

The deviation of a $y$ value from the mean $\bar{y}$ can be written:

$$(y-\bar{y})=(a+bx-\bar{y})+(y-a-bx) \tag{4.16}$$

This expresses the fact that the deviation is made up of a part which is explained by the regression of $y$ on $x$ plus the difference of the $y$ value from the prediction of the model. It breaks down the variability of $y$ into an amount due to its dependence on $x$ and a **residual** term. It is sometimes instructive to plot these residuals against $x$ or $y$ to see if the errors made by the model are cyclical or systematic in some other way. The model could then be modified to take this behaviour into account by adding more terms into it. In classical sampling theory the residuals are assumed to be normally distributed. Squaring both sides of equation (4.16), inserting the formula for $a$ as a function of $b$ and adding them together for the whole sample we obtain:

$$CS_{yy}=b^2CS_{xx}+\text{sum of squared residuals} \tag{4.17}$$

The two parts on the right-hand side of (4.16) have a product which sums to zero in this calculation and they are said to be **orthogonal**. Many statistical models have

**Table 4.5** Analysis of variance table for a linear regression model

| Source | Sum of squares | Degrees of freedom | Mean Square |
|---|---|---|---|
| Regression | $b^2CS_{xx} = r^2CS_{yy} = \dfrac{CS_{xy}^2}{CS_{xx}}$ | 1 | $\dfrac{CS_{xy}^2}{CS_{xx}}$ |
| Residual | $(1-r^2)CS_{yy} = CS_{yy} - \dfrac{CS_{xy}^2}{CS_{xx}}$ (obtained by subtraction) | $n-2$ (obtained by subtraction) | $\dfrac{\text{residual sum of squares}}{n-2}$ |
| Total | $CS_{yy}$ | $n-1$ | — |

orthogonal terms which means we can assess the contribution of each one to the total variance of $y$. We can see which sections of the model explain the most variability in the data. The implementation of (4.17) is usually tabulated as in Table 4.5.

The total number of degrees of freedom is $(n-1)$ because the total sum of squares has been corrected for the mean as discussed in Chapter 2. The regression has 1 degree of freedom as the calculation of the slope imposes a single condition on the $y$ values. Hence at this stage 2 of the original $n$ data values could be reconstructed from a knowledge of the others. It follows by subtraction that the residual sum of squares has $(n-2)$ degrees of freedom. Taking the corrected sums from Table 4.4 we can produce the ANOVA table for the sample we have (Table 4.6).

**Table 4.6** Analysis of variance table for the Heater Example

| Source | Sum of squares | Degrees of freedom | Mean square | F ratio |
|---|---|---|---|---|
| Regression | 1285.5051 | 1 | 1285.5051 | 16.41 |
| Residual | 1834.0022 −1285.5051 =548.4971 | 7 | 78.3567 | — |
| Total | 1834.0022 | 8 | — | — |

Notice how the sum of squares and number of degrees of freedom for the residual row are found by subtraction of the regression entries from the total. This is a standard trick in compiling ANOVA tables.

The $F$ ratio in the last column provides a method of testing whether the regression sum of squares accounts for a significant proportion of the total. The fraction of total variability 'mopped up' by the regression is the correlation coefficient squared. Testing the size of the coefficient is therefore equivalent to testing how much of the variability in the $y$ values can be explained by the regression model.

The ***F* test** is another way of doing this as it compares the regression variance with the residual variance. An $F$ ratio tests the equality of two population variances. It appears again in Chapters 6 and 7 in the analysis of data from more than one sample. The ratio is obtained by dividing the larger variance by the smaller and has two numbers of degrees of freedom, one for its numerator and one for its denominator.

In Table 4.6 the $F$ ratio is 16.41 and is the regression variance with 1 degree of freedom divided by the residual variance with 7 degrees of freedom. Of course, if the residual variance is bigger than the regression variance, then the regression has failed and no testing is necessary. We are thus performing a one-tail test to see if $F$ is significantly bigger than 1. From the tables on page 191 the 5% significant value for an $F$ ratio with $\nu_1$ equal to 1 and $\nu_2$ equal to 7 is 5.59 and our figure of 16.41 is abnormally high. It is also significant at 1%. We conclude that the variances are not the same and **the regression model gives a good description of the relationship between the variables.**

It was pointed out earlier that the $t$ statistic (4.9) tests both the correlation coefficient and the slope of the regression line against the null hypothesis value of zero. The $F$ ratio method is equivalent to this test because the square root of an $F$ ratio with one degree of freedom in the numerator has the $t$ distribution. Here are the three formulae we have used to test a regression:

$$r = \frac{CS_{xy}}{\sqrt{(CS_{xx}CS_{yy})}}$$

$$t = (CS_{xy} - \beta CS_{xx})\sqrt{\frac{n-2}{CS_{xx}CS_{yy} - CS_{xy}^2}}; \qquad \nu = n-2$$

$$F = \frac{CS_{xy}^2(n-2)}{CS_{xx}CS_{yy} - CS_{xy}^2}; \qquad \nu_1 = 1,\ \nu_2 = n-2 \tag{4.18}$$

The probability distribution of $r$ is complicated when the population correlation coefficient is not zero. A test of whether $r$ is significant compared with the null hypothesis value $\rho$ (Greek letter 'rho') is given by the statistic

$$z = (\tanh^{-1} r - \tanh^{-1} \rho)\sqrt{(n-3)} \tag{4.19}$$

which is approximately normally distributed. The symbol $\tanh^{-1}$ stands for the inverse hyperbolic tangent which is found on scientific calculators.

## Paired data

A popular and powerful type of experimental design is to apply two treatments either to the same group of subjects or to a set of identical 'twins'. One of the treatments could be a placebo to form a **control** on the effect being investigated. We are then seeking a significant improvement over and above that achieved with the control group. The two measurements might be made at times 'before' and 'after' some experiment is performed or procedure executed. The important feature is that the matching of the data values nullifies the effects on them of any extraneous factors in the situation. This enables us to focus attention solely on the treatment we are applying without worrying too much about these other phenomena. Clearly the subjects themselves should form a fairly homogeneous sample but the design does allow a certain variability in their nature as each pair of data values will be examined separately and only their relative sizes compared with each other.

In some circumstances the matching is achieved by applying the two treatments to different parts of the same subject. For instance coating two different leaves on the same plant with two kinds of pesticide. In view of the importance of the paired data structure we study four worked examples covering different measurement levels.

## The matched-pair $t$ test

When it is considered appropriate to adopt the normality assumption of classical sampling theory, then a $t$ test can be performed on the differences between the paired measurements. The normality assumption is that these differences are normally distributed and statistically independent of each other.

### The paint example

A paint manufacturer wants to test the power of a new flame resistant paint. He cut 8 pieces of wood in half and coated one half of each piece with ordinary paint and the other half with the new paint. The times taken for flames to penetrate through each of the 16 halves were:

| Piece of wood | *A* | *B* | *C* | *D* | *E* | *F* | *G* | *H* |
|---|---|---|---|---|---|---|---|---|
| Burning time for half with ordinary paint (minutes) | 16 | 18 | 13 | 22 | 17 | 18 | 17 | 20 |
| Burning time for half with new paint (minutes) | 19 | 23 | 17 | 19 | 15 | 21 | 20 | 21 |

We test whether the new paint produces significantly longer burning times than the ordinary paint.

*Solution* The pairs of measurements for each of the 8 items in the sample are truly matched as the halves of treated material to which they refer had exactly the same history prior to the experiment. The halves are the same age, came from the same tree and shared any chemical treatment in the past which might affect flammability. We can be confident that the differences in burning times between them is due solely to the two types of paint.

A single sample of improved burning times is obtained by subtracting corresponding data values. The improvements are +3, +5, +4, −3, −2, +3, +3 and +1. Notice that although the improvements for panels $D$ and $E$ are negative, indicating that the new paint performed worse than the ordinary paint, they could be random fluctuations and the new paint could still be effective.

The null hypothesis we test is that there is no overall improvement and the average population value is zero. We are prepared to reject this only if the sample average is so large and positive that the evidence for a longer burning time is established. From formulae (2.1), (2.3) and (2.7) the mean and standard deviation of the sample of improvements are 1.75 and 2.8661. The $t$ statistic (3.14) with zero as the null hypothesis value of $\mu$ is 1.727. This has $(8-1)$ degrees of freedom and from the tables on page 192 the significant one-tail value at 5% is 1.895. We do a one-tail test because we seek evidence of a positive $\mu$ value indicating a positive benefit from the new paint. However, as the calculated $t$ statistic fails to reach the critical value, **there is no evidence to support the effectiveness of the new paint.**

## The Wilcoxon matched-pair signed-ranks test

It may happen that although metric variables are being measured the normality assumption is not justified. If it makes sense in the context of the situation to rank the differences in the data values, then the following nonparametric test can help.

### The teaching example

Seven children were given a geography test before and after they had seen a television programme about Venezuela. Their marks were out of a maximum of 20:

| Child | Alf | Bob | Claire | Doris | Egbert | Freda | Gregory |
|---|---|---|---|---|---|---|---|
| First test mark | 14 | 13 | 9 | 16 | 15 | 19 | 16 |
| Second test mark | 17 | 17 | 8 | 20 | 20 | 19 | 13 |

Has the television programme increased their knowledge significantly?

*Solution* Differences of marks out of 20 are not on a continuous scale and, from the educational point of view, lack precision as measures of knowledge and ability. It is arguably more appropriate to use ranking methods in their analysis than, say, the matched-pair $t$ test considered above. Going to the other extreme and discarding information about relative size, we could apply the sign test to the number of improvements, pluses, out of the 7 trials. **The Wilcoxon matched-pair signed-ranks test** bridges the gap and allows us to retain the information content of ordinal data without having to make the normality assumption for it.

Firstly the absolute sizes of the differences are ranked but the ranks for positive differences kept separate from the ranks for negative ones. This is shown in Table 4.7. Differences which are the same size are given the average of the ranks they would have obtained otherwise. For instance suppose 4 items tie for 5th place. Together they occupy positions 5, 6, 7 and 8; hence each one is given the rank 6.5. In the table, Bob's and Doris' differences receive a rank of 4.5 each because they tie for positions 4 and 5. Differences of zero, like Freda's, are ignored, and the sample size reduced accordingly.

**Table 4.7** Ranked differences for the Teaching Example

| Child | Alf | Bob | Claire | Doris | Egbert | Freda | Gregory |
|---|---|---|---|---|---|---|---|
| Improvement | +3 | +4 | −1 | +4 | +5 | 0 | −3 |
| Rank of positive differences | 2.5 | 4.5 | | 4.5 | 6 | — | |
| Rank of negative differences | | | 1 | | | — | 2.5 |

The second step is to calculate the smaller of the two row sums for the ranks. In our case the ranks for negative differences add up to 3.5 and this is smaller than the sum of the positive ones. As we are looking for a significant increase in the scores we expect the negative differences to account for the smaller sum of ranks and would declare the television programme to be ineffectual if this were not the case.

In identifying the smaller of the two row totals for larger samples it is helpful to know that the sum of $m$ ranks is $m(m+1)/2$. Thus one row can be totalled and the result subtracted from $m(m+1)/2$ to obtain the total of the other row. The smaller of the two amounts is then used. Note that $m$ is the sample size less the number of differences which were not ranked because they were zero.

The final stage of the test is to determine whether the total is significant. The smaller it is the more pronounced the treatment is in dividing the sample into positive and negative differences. The table on page 194 gives one-tail critical values, and for an effective sample of size of 6 such as ours the total must be less than or equal to 2 to be significant at the 5% level. Our value of 3.5 is therefore not significant and **the television programme does not seem to have been beneficial**. The conscientious reader might care to calculate the $t$ statistic for the sample of differences. It too indicates that the average difference is not significantly bigger than zero.

A two-tail test can be carried out using the table on page 194 by simply doubling the probabilities as indicated there. Tables are not needed for effective sample sizes bigger than 16 as there is a normal approximation. The formula for this is quoted underneath the table on page 194.

## Rank correlation

The correlation coefficient defined in the previous section can be applied to ordinal data as well as to metric data. The formula can be simplified in such cases and it provides a useful measure of the level of agreement between two rankings of the same subjects. Here is an example.

### The dog show example

Six dogs are ranked by two judges in a competition:

| Dog | *A* | *B* | *C* | *D* | *E* | *F* |
|---|---|---|---|---|---|---|
| Rank given by first judge | 2 | 6 | 1 | 4 | 3 | 5 |
| Rank given by second judge | 4 | 6 | 2 | 5 | 1 | 3 |

Do the judges show a significant level of disagreement in their verdict?

*Solution* Taking one of the sets of ranks as the $x$ variable and the other as $y$ we could calculate the correlation coefficient from Equation (4.6). Alternatively, as the variables are ranks, the formula can be expressed in terms of the differences in the rankings of each subject:

$$\textbf{Spearman's rank correlation coefficient, } r_s = 1 - \frac{6(\text{sum of (difference in rank)}^2)}{n(n^2-1)} \tag{4.20}$$

For our data the differences are 2, 0, 1, 1, −2 and −2 and so the sum of their squares is 14. As $n$, the number of items in the sample, is equal to 6, the coefficient is 0.6.

For samples with 10 or less subjects being ranked the significance of the coefficient can be tested by reference to the table on page 196. For samples with more than 10 items the $t$ statistic of equation (4.9) can be evaluated and tested. Now the correlation coefficient is always between −1 and +1 and if it is large and positive then the two sets of ranks are in agreement with each other. If it is large and negative there is a systematic disagreement in which the same items receive low ranks on one scale and high ranks on the other. Values of the coefficient near to zero imply that there is no connection between the rankings at all.

We can test for the judges' level of agreement by performing a one-tail test. From the table on page 196 the significant value of the Spearman's rank correlation coefficient for a sample of 6 items is 0.829 at the 5% level. Hence **there is no evidence that the judges agree with each other about the ranking of the dogs.**

Spearman's coefficient can be used to measure the degree of correlation between two metric variables when the assumption of normality would be inappropriate. The items in the sample are arranged in order according to the value of one of the variables and assigned ranks. They are then sorted according to the other variable and receive different ranks. Ties in these ranking procedures are dealt with by the average rank method mentioned in the Teaching Example. The level of agreement of the two sets of ranks measures how well the original variables are correlated in so far as putting the sample items in sequence is concerned. This process could be applied to the data in the Teaching Example to determine whether the two tests rank the children in essentially the same order of excellence or not.

## The McNemar test

The incidence of two attribute variables can be displayed as a contingency table. As in the Drinks Example it is sometimes necessary to consider a metric variable like age as an attribute, when it has been collected as a classified response in a questionnaire, say. The standard apparatus consisting of the $\chi^2$ and binomial distributions are brought to bear and in the following example a test for the direction of a pre-treatment to post-treatment change is executed.

### The hydrangea example

A new fungicide is being tested on red and blue hydrangeas but has a tendency to make them change colour for the following season. A sample of 96 plants gave the following results:

| | | Colour after fungicide application | |
|---|---|---|---|
| | | Red | Blue |
| Colour before fungicide application | Red | 42 | 13 |
| | Blue | 16 | 25 |

The table shows that 29 plants changed colour in one way or another. If these changes could have occurred without the fungicide treatment then we might begin by investigating whether the overall number of changes is significant. If, say, 10% of hydrangeas change colour spontaneously, then the expected number out of 96 plants would be 9.6. This could be tested against the observed figure of 29 using the $\chi^2$ goodness-of-fit test. Equivalently, as explained in the Hand Cream Example of Chapter 3, the normal approximation to the binomial distribution can be applied by thinking of 'colour change' as 'success' in 96 trials, each with probability 0.1 of success. We assume either that this test is significant or that changes are not to be expected spontaneously at all. In other words we accept that the 29 changes are due to the treatment effect alone.

The table can be thought of as a contingency table and the $\chi^2$ test for an association between the classifications carried out. This simply confirms that there is a connection between the pre-treatment colour and the post-treatment colour. It does not enlighten us on the detailed nature of the changes. The question we ask can be phrased as follows: given that there were 29 colour changes all due to the fungicide, do they indicate that the changes 'red to blue' and 'blue to red' are equally likely to occur?

*Solution* **The McNemar test** concentrates attention on the cells in the table which correspond to changes in colour. The null hypothesis is that the probability of a change from red to blue is the same as that from blue to red. Given that there were 29 changes, half of them, 14.5, should have been 'red to blue' and the other half 'blue to red'. These expected frequencies are to be compared with the observations 16 and 13. The appropriate $\chi^2$ statistic incorporating Yates' correction as described in Chapter 3 is

$$\chi^2 = \frac{(16-14.5-0.5)^2}{14.5} + \frac{(14.5-13-0.5)^2}{14.5} = 0.1379 \tag{4.21}$$

As the 5% significant value of $\chi^2$ with 1 degree of freedom is 3.8415 **there is no evidence to show that the changes are occurring in a dominant direction.**

The McNemar test just described is identical with the sign test of Chapter 3. We could designate a colour change from red to blue as 'plus' and from blue to red as 'minus' and test the distribution of signs with the normal approximation to the binomial. It is also strikingly similar to the Coin Example in the same chapter.

## Summary

**1** Contingency tables of cross-tabulations show the frequencies of different combinations of two attributes. The $\chi^2$ goodness-of-fit test helps identify any association between the variables. Categories of attributes may have to be merged to increase the expected frequencies in order to make the test possible. Merging is also done to investigate specific hypotheses about groups of classifications.

**2** Two metric variables can have a negative, zero or positive correlation with each other. This can be seen on a scatter diagram. A linear regression analysis attempts to describe their relationship by a model whose graph is a straight line. The success of the regression is measured by the correlation coefficient, a $t$ statistic for the slope or an $F$ ratio for the amount of variability explained by the model. The $F$ ratio calculation can be set out as an analysis of variance table. All three methods are mathematically equivalent. Point predictions and prediction intervals for the dependent variable can be derived from a regression analysis as well as confidence intervals for population means when the independent variable is given a fixed value. If a linear model is inappropriate the variables can be transformed, possibly by trial and error, in an attempt to linearise their relationship.

**3** When two variables are matched or paired in some way then various single-sample techniques can be applied to their differences. Matching is a desirable property of an experimental design as it negates the effect of factors extraneous to the treatment under study. The $t$ test is used for metric variables, the Wilcoxon matched-pair signed-ranks test for ordinal data or ranked metric data, and the $\chi^2$, binomial and McNemar tests for attribute data.

## Further reading

The $\chi^2$ distribution appears in the first part of this chapter as the principal tool for testing the observed and expected frequencies in contingency tables. In view of the Central Limit Theorem these tests are considered to be virtually distribution free. Apart from being described in most books on general statistics they occur in all of those listed in the 'Nonparametric Methods' section of the Bibliography. Reference (32) is devoted exclusively to the $\chi^2$ distribution and develops the analysis of contingency tables into partitioning, trend detection and applications with more than two variables. Fisher's exact test for $2 \times 2$ tables with small frequencies is also explained there.

Reference (35) is dedicated to regression analysis although all general statistics texts cover it with varying degrees of sophistication. The fitting of polynomial functions to data and other specific applications of linear and non-linear regression can be found in (9). A time series as defined in Chapter 2 is a single sample with two variables. Specialised texts like (60) and (65) use, amongst other techniques, regression and autocorrelation, the correlation of values with earlier values, to model their behaviour. The subject of econometrics, dealt with in books like (36) and (59), is largely about fitting regression equations to economic time series in order to construct models of company activity or the economies of countries.

The work in this chapter on paired measurements will be developed in Chapter 7 when data from more than two samples is analysed. The concept of blocking groups of experimental units is a powerful extension of the basic idea.

## Exercises

**1** Some crime statistics for three completely fictitious countries are:

| Thousands of crimes in 1984 | Burglaria | Conland | Murdia |
|---|---|---|---|
| Arson | 52 | 97 | 81 |
| Blackmail | 4 | 14 | 6 |
| Housebreaking | 72 | 132 | 110 |
| Manslaughter | 2 | 3 | 5 |
| Mugging | 132 | 264 | 219 |
| Murder | 3 | 5 | 9 |
| Shoplifting | 112 | 203 | 154 |
| Traffic offences | 216 | 450 | 300 |

(i) Combine the 'murder' and 'manslaughter' categories and test whether there is an association between the type of crime committed and the country. Why is it preferable to merge the categories for this test? (ii) Are the total numbers of crimes committed in each of the three countries consistent with the fact that Burglaria and Murdia have about the same population sizes while Conland has about twice as many people as either of them? (iii) Are the numbers of traffic offences in the three countries consistent with the population size ratios given in part (ii)?

**2** The percentage increase in weight of 12 baby floobies during their first 30 days of life were measured together with their birth weights:

| Birth weight (g) | 72 | 111 | 102 | 79 | 115 | 98 |
|---|---|---|---|---|---|---|
| Percentage weight increase in first 30 days of life | 95 | 88 | 86 | 101 | 93 | 90 |

| Birth weight (g) | 80 | 85 | 101 | 95 | 102 | 114 |
|---|---|---|---|---|---|---|
| Percentage weight increase in first 30 days of life | 100 | 88 | 89 | 91 | 89 | 82 |

(i) Plot a scatter diagram of the percentage increase against the birth weights. (ii) Calculate the regression equation of percentage weight gain on birth weight. (iii) Determine the correlation coefficient and test its value for significance. (iv) Calculate a 90% prediction interval for the percentage weight gain of a flooby who weighed 99 g at birth. (v) Evaluate a 95% confidence interval for the average weight of all floobies with birth weight 86 g.

**3** Seven carpenters were observed doing similar jobs in the morning and in the afternoon. The times taken were:

| Carpenter | *A* | *B* | *C* | *D* | *E* | *F* | *G* |
|---|---|---|---|---|---|---|---|
| Morning job time (minutes) | 20 | 22 | 19 | 21 | 20 | 23 | 23 |
| Afternoon job time (minutes) | 27 | 22 | 18 | 25 | 26 | 24 | 26 |

(i) Calculate the mean and standard deviation of all the job times. (ii) Calculate a 95% confidence interval for the population mean of all job times. (iii) Test the hypothesis that it takes a carpenter 21 minutes on average to do the job in question. (iv) Test the hypothesis that the afternoon job times are significantly longer than the morning ones using the matched-pair $t$ test. (v) Test the hypothesis of part (iv) using the Wilcoxon matched-pair signed-ranks test. (vi) Test the hypothesis of part (iv) using the sign test of Chapter 3.

**4** If the stresses on a suspension bridge are evenly distributed then its shape should be a catenary. This has the equation $y = A \cosh x + B$, where $y$ is the height of the bridge at distance $x$ from the centre of the span. Measurements taken from a model of the bridge were:

| $x$ (metres) | −4 | −3 | −2 | −1 | 0 | +1 | +2 | +3 | +4 |
|---|---|---|---|---|---|---|---|---|---|
| $y$ (metres) | 5.0 | 2.7 | 1.5 | 0.8 | 0.7 | 1.0 | 1.6 | 2.2 | 6.2 |

By transforming the $x$ variables using the cosh function, perform a linear regression to determine whether the shape is in fact a catenary. If it is, estimate the values of $A$ and $B$ in its equation.

# 5 One sample with several variables

Measuring several variables for each item in a sample produces **multivariate** data. The analysis often begins with an attempt to classify cases or variables as a way of describing their relationship with each other and as an indicator of further analysis possibilities. In plant and animal taxonomies the classification system itself is the objective of the study and cases are assigned to species according to the set of attributes they possess. In the practice of medicine, doctors classify patients into types of illness by examining certain key variables called symptoms.

To establish a classification system variables conveying the same or similar information as each other must be grouped together. They are then replaced either by a single representative member of their group or by a combination of their values, a new variable, containing their joint information. Cases can also be clustered together and possibly replaced by typical specimens for subsequent analysis. Many multivariate techniques are about reducing the number of cases or variables by grouping them together or discriminating between them.

For attribute data it is sometimes feasible to draw up many-way contingency tables and extend the methodology of Chapter 4. For metric variables the concept of a regression of one variable on another introduced in the same chapter can be developed to a dependence of one variable on several other variables.

There are problems with multivariate analysis which seem to plague most applications. It is important in studies with many variables to plan for the analysis stage at the very beginning of the operation. Sometimes the most crucial variables are not identified and hence not measured. Often the nature and quality of the data varies from one variable to another. On a questionnaire, for example, the responses to 'are you male or female?' will be more accurate than those to 'how many times did you cross a road last Wednesday?' Yet, unless we are careful, all variables are given the same weight in the analysis. It is common practice to standardise all metric variables into what we called $z$ scores in Chapter 3.

The amount of calculation involved in even a modest multivariate analysis prompts most statisticians to use a computer. There is an element of data processing in many applications, such as collating medical case studies and processing marketing research questionnaires or insurance company claim forms for trends and patterns. As well as a large number of cases, these applications have large numbers of variables. For this reason the calculations required in this chapter are reconsidered in Chapter 11 on the computing aspects of statistics. In particular the use of matrices will be avoided in this chapter for the sake of those readers unfamiliar with them but will be incorporated in Chapter 11.

Multivariate analysis is an art and not a science. Computer packages allow a statistician to play with the data, transforming or omitting some variables and giving others extra weight, until a pattern of interdependence or causality emerges. There are no rules by which this trial-and-error procedure can be optimized. Unfortunately the relationships between cases or variables which have the most statistical significance often turn out to be statements of the blindingly obvious. Conclusions like '92% of people who buy petrol have one or more vehicles' are quite acceptable to a computer hungry for results. More subtle features of the data tend to be swamped out by these gross patterns. In this chapter we consider some techniques for classifying cases and variables, the combining of variables into principal components and multiple regression.

## Cluster analysis

In order to determine the positions of cases or variables relative to each other some sort of distance measure is needed. In genetics the distance between two species can be defined as the number of attributes, called characters, they have in common. Conversely, the number of characters they do not share can be thought of as a mutation distance. Generally the situation is most difficult for nominal and ordinal variables, although even for metric ones the differences in scale and importance mentioned earlier cause problems. The main ideas and pitfalls are illustrated in the following example.

### The hotels example

Data on 8 fictitious hotels are given below:

| Hotel | Number of employees ($x_1$) | Number of bedrooms ($x_2$) | Number of parking places ($x_3$) | Swimming pool ($x_4$) | Dogs permitted ($x_5$) |
|---|---|---|---|---|---|
| Averest | 10 | 37 | 50 | no | yes |
| Bedsargood | 5 | 20 | 15 | no | no |
| Comeagain | 93 | 150 | 70 | yes | yes |
| Dustea | 25 | 90 | 20 | yes | no |
| Empyre | 6 | 19 | 4 | no | no |
| Fallover | 31 | 48 | 45 | no | yes |
| Garish | 67 | 138 | 68 | yes | yes |
| Hilltown | 89 | 142 | 94 | yes | yes |

*Question 1* Examine the relationships between the variables to decide if any can be omitted from further analysis.

*Solution* A scatter diagram shows whether a relationship exists between two variables. For more than two variables there are correspondingly more axes and the diagram is multi-dimensional. We can draw **case profiles** in just two dimensions, and Fig. 5.1 relates to the hotel data. The letters in the diagram stand for the hotel names. The number of times the case lines cross each other indicates the lack of agreement between the variables in sorting the cases into rank order. The diagram can be redrawn to minimise the number of crossings with the axes arranged in a different sequence or with some of them reversed in direction to allow for negative correlation.

Looking at the profiles, it seems that there is a high correlation between the number of parking places, $x_3$, and the permitting of dogs, $x_5$, as cases $A$, $F$ and $D$ do not cross over if the $x_4$ and $x_5$ axes are interchanged. Similarly $x_4$ correlates well with $x_2$ as all hotels above $D$ have a 'yes' value for $x_4$ while $D$ and those below it have 'no'. **We shall therefore omit the two attribute variables $x_4$ and $x_5$ from the subsequent analysis as they are correlated with other variables.** As often happens, the more sensitive level of measurement provided by metric variables has dominated the diagram.

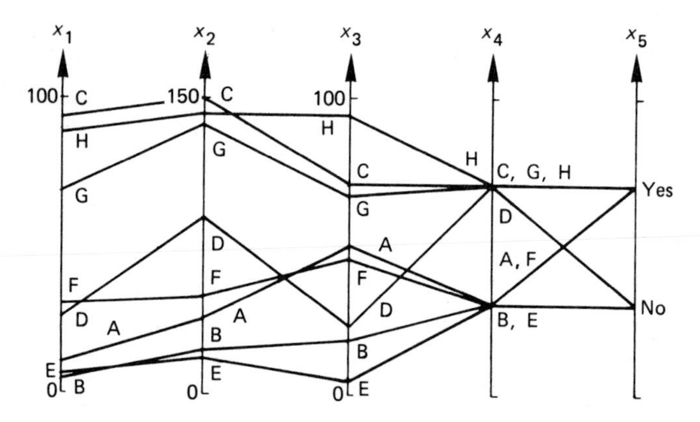


**Fig. 5.1** Case profiles for the Hotels Example

Apart from hotel *D*, which has an untypically high $x_2$ value for a hotel with small $x_1$ and $x_3$ values, the cases form two clusters, *C*, *H* and *G* in one set and *F*, *A*, *E* and *B* in another. In fact all three variables are correlated with each other and this could be examined by looking at their correlation coefficients when they are taken in pairs. Other methods of exploring bivariate correlation, like contingency tables and Spearman's coefficient, can be used on multivariate data by considering them two at a time. In the next example there are eight variables and the correlation matrix approach is demonstrated. We continue here by keeping all three variables in the working.

*Question 2* Group the hotels into clusters of similar cases.

*Solution* The hotels can be thought of as points in a three-dimensional space with coordinates $z_1$, $z_2$ and $z_3$. These are the **z scores** for each of the variables $x_1$, $x_2$ and $x_3$ as defined by equation (3.1) with $\bar{x}$ and $s$ as estimates of $\mu$ and $\sigma$:

$$z = \frac{x - \bar{x}}{s} \tag{5.1}$$

A $z$ score measures the deviation of an $x$ value from the average $\bar{x}$ in units of the standard deviation. It is preferable to work with $z$ scores as they all have the same variance, 1, and the same mean, 0. Scores on different variables are therefore comparable. Conversion of data to $z$ scores is common practice in a multivariate analysis.

Visualising each hotel as a point in three-dimensional space leads us to define the geometric distance between points as a **distance function** between hotels. Geometric or **Euclidean distance** is

$$(\text{distance})^2 = (\text{difference in } z_1 \text{ values})^2 + (\text{difference in } z_2 \text{ values})^2 \\ (\text{difference in } z_3 \text{ values})^2 + \cdots \tag{5.2}$$

or, in terms of the original $x$ variables,

$$(\text{distance})^2 = \frac{(\text{difference in } x_1 \text{ values})^2}{\text{variance of } x_1} + \frac{(\text{difference in } x_2 \text{ values})^2}{\text{variance of } x_2} \\ + \frac{(\text{difference in } x_3 \text{ values})^2}{\text{variance of } x_3} + \cdots \tag{5.3}$$

As an example of the use of this equation we calculate the distance between Comeagain and Garish:

$$(\text{distance})^2 = \frac{(93-67)^2}{1360} + \frac{(150-138)^2}{3200} + \frac{(70-68)^2}{969}$$

$$= 0.5462 \qquad (5.4)$$

Hence the distance is $\sqrt{0.5462}$, which is 0.74. The variances and result of equation (5.4) are often rounded off quite ruthlessly as only the relative sizes of the distances are important. Table 5.1 shows the distances between all possible pairs of hotels as a matrix. Generating such matrices by computer is discussed in Chapter 11.

**Table 5.1** Distance matrix for the Hotels Example

| | Averest | Bedsargood | Comeagain | Dustea | Empyre | Fallover | Garish | Hilltown |
|---|---|---|---|---|---|---|---|---|
| Averest | 0.00 | 1.17 | 3.08 | 1.40 | 1.52 | 0.62 | 2.43 | 3.17 |
| Bedsargood | 1.17 | 0.00 | 3.75 | 1.36 | 0.35 | 1.29 | 3.17 | 4.03 |
| Comeagain | 3.08 | 3.75 | 0.00 | 2.67 | 3.93 | 2.59 | 0.74 | 0.79 |
| Dustea | 1.40 | 1.36 | 2.67 | 0.00 | 1.45 | 1.11 | 2.10 | 3.08 |
| Empyre | 1.52 | 0.35 | 3.93 | 1.45 | 0.00 | 1.57 | 3.37 | 4.26 |
| Fallover | 0.62 | 1.29 | 2.59 | 1.11 | 1.57 | 0.00 | 2.01 | 2.78 |
| Garish | 2.43 | 3.17 | 0.74 | 2.10 | 3.37 | 2.01 | 0.00 | 1.03 |
| Hilltown | 3.17 | 4.03 | 0.79 | 3.08 | 4.26 | 2.78 | 1.03 | 0.00 |

One way of clustering items based on their distances apart is to draw a **minimum** or **minimal spanning tree**. This is a diagram showing how each case is related to the others in terms of the shortest distances in the table. The concept can also be used to display the relationship between variables instead of cases. The correlation coefficient can be interpreted as a measure of distance for them and is employed as such in the next example.

The algorithm for deriving minimal spanning trees is borrowed from operational research, where such trees are drawn to solve problems like wiring several houses together with the least possible amount of electrical cable. The method is as follows:

*Step 1 Choose any case or variable and join it to the case or variable nearest to it.* In our example we could start with the Empyre hotel. According to Table 5.1 its nearest neighbour is the Bedsargood and so we draw them with a line linking them together.

*Step 2 Choose the case or variable closest to those already drawn and add it to the tree.* The nearest of all the remaining hotels to the Empyre or the Bedsargood is the Averest and so adding it to the tree we obtain Fig. 5.2.

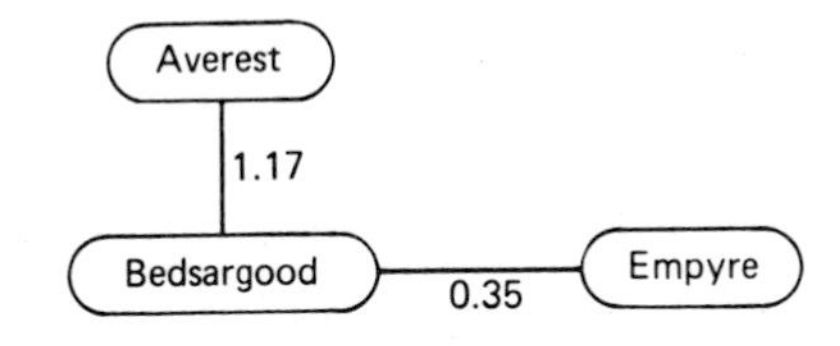


**Fig. 5.2** Beginning the minimal spanning tree

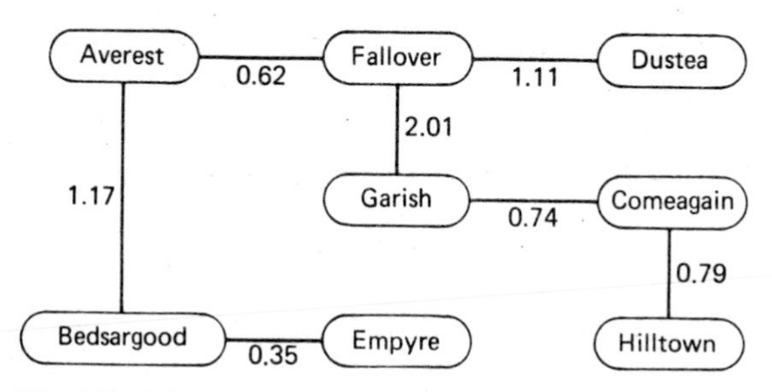


**Fig. 5.3** Minimal spanning tree for the Hotels Example

*Step 3 Repeat step 2 until all the cases or variables are included in the tree.* The result is shown in Fig. 5.3.

Having constructed a minimal spanning tree we can delete links representing the largest distances to leave clusters of similar cases. The biggest distance in Fig. 5.3 is between Fallover and Garish with the second biggest being Averest and Bedsargood. Deleting these links produces three clusters of hotels, (a) the Bedsargood and the Empyre, (b) the Averest, the Fallover and the Dustea, and (c) the Garish, the Comeagain and the Hilltown. It is comforting to see the same grouping emerge as in the profiles analysis of Fig. 5.1.

Minimal spanning trees create **single linkage clusters** because an item belongs to a cluster if it has just one link into it. The item does not have to relate to the 'centre' of the cluster in any sense but only to the 'edge'. Single linkage clusters tend to be long and sausage-shaped for this reason.

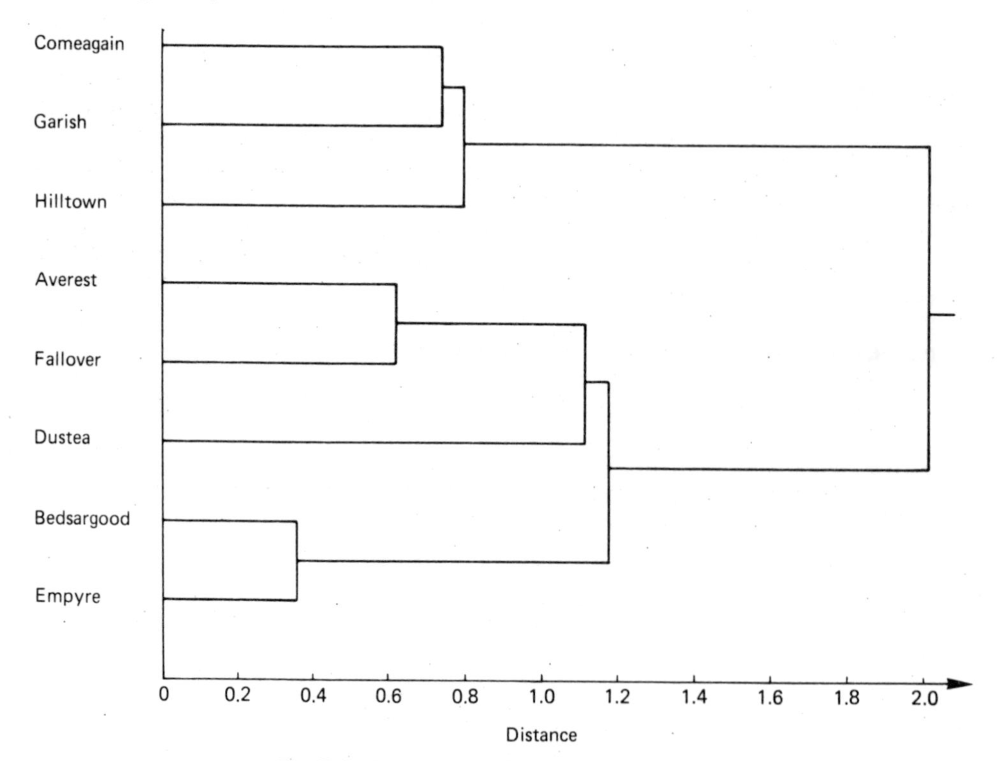


**Fig. 5.4** Dendrogram for the Hotels Example

The process of deleting successively smaller links in a minimal spanning tree can be illustrated by a **dendrogram** or **tree diagram** as in Fig. 5.4. It shows how the cases cluster together as the nearest neighbour threshold distance is increased. **Either Fig. 5.3 or Fig. 5.4 or both form the solution to Question 2.**

*Question 3* Which variable acts as the best predictor of the cluster a hotel belongs to?

*Solution* We look for a numerical property of a hotel whose value is correlated with the cluster to which the hotel belongs. This is a simple form of **discriminant analysis.** In general a **discriminant function** of several variables is sought which behaves in some predictive or diagnostic way. In this example the ranking of the hotels according to each of the three variables can be seen in Fig. 5.1. Comparing it with the dendrogram, Fig. 5.4, it appears that $x_3$ is the best predictor of relative position in the clustering. **Hence the number of parking places a hotel has is a good indicator of its overall characteristics.**

## Principal components

Multivariate data provides information about the relationships between the variables being measured as well as about the subjects in the sample. Besides clustering the variables as mentioned earlier we can search for a combination of them which as a single measurement accounts for a large proportion of the total variability in the sample. The linear combination corresponding to the largest amount of variability is called the **first principal component.** The **second principal component** accounts for the most variability after the effect of the first has been removed. This process continues and in theory there are as many principal components as there are variables. In practice the first few usually explain a significant proportion of the original total variance and the remainder can be ignored. The analysis attempts to transform the set of variables into a smaller set of derived measurements which describe salient features of the sample.

An early application of principal components was to measurements made on the bodies of criminals. The purpose was to enable a person to be identified as the technique of photography was yet to be invented. The analysis was successful in reducing the number of variables to one component describing head size, one describing length of limbs and so on. Each component brings together correlated variables and expresses their information content as a single value.

### The psychiatry example

Ten patients were given a variety of tests to assess 8 aspects of their behaviour. The results of these fictitious experiments were:

| Patient | *A* | *B* | *C* | *D* | *E* | *F* | *G* | *H* | *I* | *J* |
|---|---|---|---|---|---|---|---|---|---|---|
| Anxiety ($x_1$) | 4.8 | 4.9 | 4.7 | 3.6 | 3.6 | 5.2 | 2.1 | 4.0 | 2.5 | 2.0 |
| Delusions of grandeur ($x_2$) | 0.4 | 1.2 | 0.5 | 1.5 | 4.2 | 2.5 | 4.3 | 2.1 | 4.3 | 4.3 |
| Depression ($x_3$) | 5.5 | 4.3 | 5.9 | 4.5 | 5.3 | 1.6 | 0.9 | 2.8 | 0.8 | 0.6 |
| Hallucinations ($x_4$) | 1.5 | 1.6 | 1.4 | 2.1 | 1.8 | 1.6 | 1.7 | 3.3 | 1.5 | 1.6 |
| Hostility ($x_5$) | 2.2 | 2.5 | 2.2 | 3.7 | 3.4 | 2.8 | 4.2 | 2.9 | 4.6 | 4.5 |
| Motor abnormality ($x_6$) | 4.5 | 4.4 | 4.3 | 2.3 | 4.0 | 1.4 | 0.3 | 1.6 | 0.5 | 0.2 |
| Speech disorder ($x_7$) | 1.8 | 1.9 | 1.8 | 1.6 | 1.9 | 2.9 | 1.6 | 2.3 | 1.2 | 1.6 |
| Thought disorder ($x_8$) | 3.1 | 2.8 | 2.6 | 3.0 | 4.2 | 3.0 | 3.8 | 4.2 | 3.5 | 3.8 |

*Question 1* Derive a clustering scheme for the variables being measured.

*Solution* As well as grouping the patients, as we did the hotels in the last example, it is equally interesting here to investigate how the variables, the different behaviour patterns, are related to each other. To form clusters we need a distance function and for metric variables it is common to base this on the correlation coefficient defined in Chapter 4. The expression:

$$\text{Distance between two variables} = 1 - \text{correlation coefficient} \tag{5.5}$$

produces distances which are small for highly positively correlated variables which are conveying similar information and large for those which are negatively correlated. The correlation coefficients between all possible pairs of variables can be displayed as a **correlation matrix**:

**Table 5.2** Correlation matrix for the Psychiatry Example

| | $x_1$ | $x_2$ | $x_3$ | $x_4$ | $x_5$ | $x_6$ | $x_7$ | $x_8$ |
|---|---|---|---|---|---|---|---|---|
| $x_1$ | 1.00 | −0.81 | 0.64 | 0.00 | −0.93 | 0.73 | 0.66 | −0.59 |
| $x_2$ | −0.81 | 1.00 | −0.72 | −0.04 | 0.86 | −0.71 | −0.26 | 0.72 |
| $x_3$ | 0.64 | −0.72 | 1.00 | −0.04 | −0.74 | 0.95 | 0.06 | −0.38 |
| $x_4$ | 0.00 | −0.04 | −0.04 | 1.00 | −0.04 | −0.18 | 0.29 | 0.53 |
| $x_5$ | −0.93 | 0.86 | −0.74 | −0.04 | 1.00 | −0.81 | −0.55 | 0.51 |
| $x_6$ | 0.73 | −0.71 | 0.95 | −0.18 | −0.81 | 1.00 | 0.12 | −0.45 |
| $x_7$ | 0.66 | −0.26 | 0.06 | 0.29 | −0.55 | 0.12 | 1.00 | −0.07 |
| $x_8$ | −0.59 | 0.72 | −0.38 | 0.53 | 0.51 | −0.45 | −0.07 | 1.00 |

Correlation matrices are readily obtainable from computer statistics packages although equation (4.6) could be applied to every possible pair of variables by hand. It should be remembered that the correlation coefficient measures linear correlation only and also that spurious or artificial correlation will emerge if two variables are both correlated to a third variable. The next question in this example concerns this phenomenon. We proceed here to construct the distance matrix using (5.5):

**Table 5.3** Distance matrix for the Psychiatry Example variables

| | $x_1$ | $x_2$ | $x_3$ | $x_4$ | $x_5$ | $x_6$ | $x_7$ | $x_8$ |
|---|---|---|---|---|---|---|---|---|
| $x_1$ | 0.00 | 1.81 | 0.36 | 1.00 | 1.93 | 0.27 | 0.34 | 1.59 |
| $x_2$ | 1.81 | 0.00 | 1.72 | 1.04 | 0.14 | 1.71 | 1.26 | 0.28 |
| $x_3$ | 0.36 | 1.72 | 0.00 | 1.04 | 1.74 | 0.05 | 0.94 | 1.38 |
| $x_4$ | 1.00 | 1.04 | 1.04 | 0.00 | 1.04 | 1.18 | 0.71 | 0.47 |
| $x_5$ | 1.93 | 0.14 | 1.74 | 1.04 | 0.00 | 1.81 | 1.55 | 0.49 |
| $x_6$ | 0.27 | 1.71 | 0.05 | 1.18 | 1.81 | 0.00 | 0.88 | 1.45 |
| $x_7$ | 0.34 | 1.26 | 0.94 | 0.71 | 1.55 | 0.88 | 0.00 | 1.07 |
| $x_8$ | 1.59 | 0.28 | 1.38 | 0.47 | 0.49 | 1.45 | 1.07 | 0.00 |

The calculations in Tables 5.2 and 5.3 have been rounded off to 2 decimal places for clarity. Just as in the Hotels Example, a minimal spanning tree can be generated from the distance matrix and is shown in Fig. 5.5. A dendrogram could be drawn if required. **Figure 5.5 shows that depression, motor abnormality, anxiety and speech disorder form**

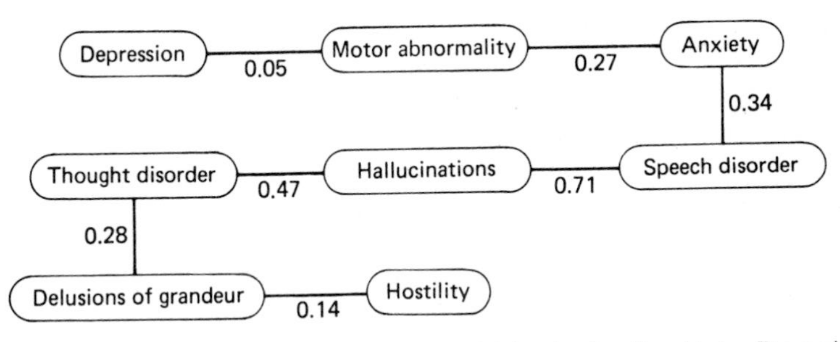


**Fig. 5.5** Minimal spanning tree for the variables in the Psychiatry Example

**one cluster, hallucinations forms a cluster on its own and thought disorder, delusions of grandeur and hostility make up a third.**

*Question 2* It seems from the minimal spanning tree that the variables anxiety, depression and motor abnormality form a cluster. To what extent is the correlation between anxiety and depression a consequence of their joint statistical dependence on motor abnormality?

*Solution* It is sometimes helpful in investigating spurious correlation to calculate a **partial correlation coefficient**. This is the residual correlation when the dependence of two variables $x$ and $y$ on a third variable $z$ has been removed by linear regression. The formula is:

$$\text{partial correlation coefficient between } x \text{ and } y \text{ allowing for the effect of } z = \frac{r_{xy} - r_{xz}r_{yz}}{\sqrt{[(1-r_{xz}^2)(1-r_{yz}^2)]}} \quad (5.6)$$

Applying it to the three variables mentioned in the question:

$$\begin{aligned}\text{partial correlation between anxiety and depression allowing for the effect of motor abnormality} &= \frac{0.64-(0.73)(0.95)}{\sqrt{[(1-0.73^2)(1-0.95^2)]}} \\ &= -0.25 \quad (5.7)\end{aligned}$$

As with many multivariate statistics, it is difficult to interpret this in the context of the original situation. **It tells us that once those parts of anxiety and depression explained by motor abnormality levels are removed from their measurement, the remainders are uncorrelated and, if anything, act as negative indicators of each other.**

*Question 3* Is there a combination of variables which can be used to explain a large proportion of the variability observed amongst the patients?

*Solution* Correlation coefficients, distance functions, minimal spanning trees and dendrograms comprise a method of clustering variables in the same way as they provide a method for clustering cases. For variables, however, we can extend the analysis and attempt to create new quantities which are more descriptive in some sense than the existing ones. The determination of principal components, which were discussed earlier, requires the **eigenvalues** and **eigenvectors** of the correlation matrix. These are mathematical objects associated with a matrix, and their origins lie in the algebra of linear transformations. Their definition and calculation are discussed in Chapter 11 as in practice a computer is almost essential for their evaluation. Statistics packages which cater for multivariate analysis include their determination as part of their repertoire.

The use of the correlation matrix has the effect of analysing the $z$ scores rather than the original raw data values. This is usually preferable, for the same reason that the $z$ scores were taken in the Hotels Example; each variable is reduced to the same mean and variance and so holds equal importance in the calculations.

The eigenvalues for the correlation matrix (5.2) are 4.72, 1.61, 0.99, 0.51, 0.09, 0.05, 0.02 and 0.01. In general there are as many eigenvalues as variables and in most practical applications they decrease rapidly from largest to smallest, as they do here. The total of all the eigenvalues should be equal to the number of variables and the fraction of each one to the total measures how much of the overall variability its eigenvector accounts for. Dividing each eigenvalue by 8 we obtain 0.59, 0.20, 0.12, 0.06, 0.01 and 0.01 with the last two being zero to 2 decimal places. Hence the first three eigenvalues account for 0.91 of the total variability and their eigenvectors which are the principal components, are given in Table 5.4. They were obtained from a computer package as described in Chapter 11.

**Table 5.4** First 3 principal components for the Psychiatry Example

| | First principal component | Second principal component | Third principal component |
|---|---|---|---|
| Eigenvalue | 4.72 | 1.61 | 0.99 |
| Percentage of total variability accounted for | 59 | 20 | 12 |
| Correlation with: | | | |
| Anxiety | −0.94 | −0.22 | −0.22 |
| Delusions of grandeur | 0.91 | −0.02 | −0.02 |
| Depression | −0.83 | 0.12 | 0.50 |
| Hallucinations | 0.09 | −0.85 | 0.37 |
| Hostility | 0.96 | 0.19 | 0.02 |
| Motor abnormality | −0.88 | 0.17 | 0.36 |
| Speech disorder | −0.42 | −0.71 | −0.50 |
| Thought disorder | 0.67 | −0.51 | 0.43 |

The coordinates of each principal component along an axis representing a variable is the correlation coefficient of the component with the variable. Hence the first component is associated with delusions of grandeur and hostility with negative relationships with anxiety, depression and motor abnormality. The second component is an 'anti' hallucinations and speech disorder one, while the third is not really significant in any of its correlations. **We conclude that a measurement associated with delusions of grandeur, hostility, anxiety, depression and motor abnormality dominates, while a secondary measure based on hallucinations and speech disorder could be used.**

The information contained in the identification of the principal components may in itself form an adequate analysis of the data. The coefficients in Table 5.4 could be applied to the $z$ scores of the patients to derive their ratings as measured by each component. The cases could then be ranked according to the ratings in an attempt to cluster them. Alternatively the two or three new variables we have created as linear combinations of the original ones could be taken as replacements for them in future experiments or assessments of patients.

## Multivariate regression

A **multivariate or multiple regression** is the fitting of the model

$$y = \alpha + \beta_1 x_1 + \beta_2 x_2 + \beta_3 x_3 + \cdots + \beta_p x_p + \text{error} \qquad (5.8)$$

as a prediction of the dependent variable $y$ on the independent variables $x_1$, $x_2$, $x_3, \ldots, x_p$. The rationale behind such a model is an extension of that given in Chapter 4 for one independent variable. In the multivariate case the geometrical representation of the model is a plane in $(p+1)$-dimensional space and so the assistance of scatter diagrams and graphs is not available to us as it was there. Two examples of multiple regression follow.

## The girls' weights example

The weights of 8 girls were measured together with their heights and ages.

| Girl | Weight (kg) $y$ | Height (cm) $x_1$ | Age (years) $x_2$ |
|---|---|---|---|
| Alice | 29 | 146 | 9 |
| Bertha | 35 | 138 | 10 |
| Clare | 31 | 143 | 9 |
| Denise | 28 | 127 | 8 |
| Elouise | 26 | 141 | 9 |
| Freda | 37 | 132 | 10 |
| Gertie | 34 | 155 | 12 |
| Helen | 26 | 128 | 7 |

*Question 1* Fit the regression model

$$y = \alpha + \beta_1 x_1 + \beta_2 x_2 + \text{error}$$

in order to be able to predict the weight of a similar girl from a knowledge of her height and age.

*Solution* The quantities $\alpha$, $\beta_1$ and $\beta_2$ in the model are chosen to give the most accurate predictions possible for the sample. It is then hoped that they will give acceptable results for other members of the population. The methods of calculus enable us to derive a technique for calculating estimates $a$, $b_1$ and $b_2$ for them which produce the least amount of total squared prediction errors. This is known as the **method of least squares** and generates the following set of **normal equations**:

$$\begin{aligned} CS_{x_1x_1}b_1 + CS_{x_1x_2}b_2 &= CS_{x_1y} \\ CS_{x_2x_1}b_1 + CS_{x_2x_2}b_2 &= CS_{x_2y} \\ a &= \bar{y} - b_1\bar{x}_1 - b_2\bar{x}_2 \end{aligned} \qquad (5.9)$$

The letters $CS$ stand for 'corrected sums' and were defined in Chapter 4. When there are more than two independent variables these equations are extended in the obvious way (see equations (11.5)). Matrix and computer techniques for solving the normal equations are dealt with in Chapter 11. To help the reader appreciate the nature of

**Table 5.5** Sums of squares and products for the Girls' Weights Example

| Girl | $y$ | $x_1$ | $x_2$ | $y^2$ | $x_1^2$ | $x_2^2$ | $yx_1$ | $yx_2$ | $x_1x_2$ |
|---|---|---|---|---|---|---|---|---|---|
| Alice | 29 | 146 | 9 | 841 | 21 316 | 81 | 4 234 | 261 | 1 314 |
| Bertha | 35 | 138 | 10 | 1 225 | 19 044 | 100 | 4 830 | 350 | 1 380 |
| Clare | 31 | 143 | 9 | 961 | 20 449 | 81 | 4 433 | 279 | 1 287 |
| Denise | 28 | 127 | 8 | 784 | 16 129 | 64 | 3 556 | 224 | 1 016 |
| Elouise | 26 | 141 | 9 | 676 | 19 881 | 81 | 3 666 | 234 | 1 269 |
| Freda | 37 | 132 | 10 | 1 369 | 17 424 | 100 | 4 884 | 370 | 1 320 |
| Gertie | 34 | 155 | 12 | 1 156 | 24 025 | 144 | 5 270 | 408 | 1 860 |
| Helen | 26 | 128 | 7 | 676 | 16 384 | 49 | 3 328 | 182 | 896 |
| Total | 246 | 1 110 | 74 | 7 688 | 154 652 | 700 | 34 201 | 2 308 | 10 342 |
| Mean | 30.75 | 138.75 | 9.25 | | | | | | |
| Corrected sum | | | | 123.5 | 639.5 | 15.5 | 68.5 | 32.5 | 74.5 |

the calculations we proceed here by hand and Table 5.5 shows the derivation of the various summations required.

The corrected sums in this table are formed in the same way as in Table 4.3. Substituting values from Table 5.5 into the normal equations (5.9) we obtain:

$$639.5b_1 + 74.5b_2 = 68.5$$

$$74.5b_1 + 15.5b_2 = 32.5 \qquad (5.10)$$

$$a = 30.75 - 138.75b_1 - 9.25b_2$$

The simultaneous equations in the $b$s can be solved by systematically eliminating each unknown in turn. The equation with the numerically largest coefficient is divided by that number. Our first equation becomes:

$$b_1 + 0.1165b_2 = 0.1071 \qquad (5.11)$$

We now subtract an appropriate multiple of this equation from the second equation to eliminate $b_1$. Multiplying it by 74.5 and subtracting from the second equation:

$$6.821b_2 = 24.521 \qquad (5.12)$$

If there were more than two $b$ values and hence more equations, the unknown $b_1$ could be eliminated from all of them in this manner. The procedure could be repeated to remove another $b$ value and continued until just one remains, as we have now in equation (5.12). We solve this to find that:

$$b_2 = 24.521/6.821 = 3.595 \qquad (5.13)$$

Back-substituting this answer provides the other $b$ value from equation (5.11):

$$b_1 = 0.1071 - (0.1165)(3.595) = -0.312 \qquad (5.14)$$

Finally, $a$ is given by the last equation of (5.10) to be 40.8 and so **the regression model is**

$$\mathbf{y = 40.8 - 0.3x_1 + 3.6x_2}$$

A word of warning should be said about accuracy here. When some of the variables are strongly correlated with each other there is an almost linear relationship between the coefficients in the normal equations. This is disastrous from the point of view of numerical stability and the equations are **ill-conditioned**. Their solution involves dividing by very small numbers which, if they have been rounded off excessively, produce inaccurate results. The system of equations effectively degenerates to a smaller one as those which are merely combinations of other equations do not represent useful additional information at all. The problem of **multicollinearity** is extremely common in multivariate analysis as the very statistical properties we are seeking, linear relationships between variables, render the solution of our equations and the determination of eigenvalues and eigenvectors difficult and unreliable. It is advisable to discard variables which are highly correlated with other variables before starting a regression or principal components analysis. The correlation matrix can be examined to identify the offending variables for this purpose.

*Question 2* Test the validity of the regression model as a description of the data.

*Solution* The correlation coefficient defined in Chapter 4 between two variables measures the degree to which a straight-line graph describes their relationship. The **multiple correlation coefficient or multiple regression coefficient, *R***, is the two-variable correlation coefficient between the predicted $y$ values of a multiple regression model and the observed $y$ values in the sample. A convenient formula for its square is

$$R^2 = \frac{\text{variance of predicted } y \text{ values}}{\text{variance of observed } y \text{ values}}$$

$$= \frac{b_1 CS_{yx_1} + b_2 CS_{yx_2} + b_3 CS_{yx_3} + \cdots + b_p CS_{yx_p}}{CS_{yy}}. \qquad (5.15)$$

For the data we have the corrected sums from Table (5.5) and the $b$ values found give:

$$R^2 = \frac{-0.3 \times 68.5 + 3.6 \times 32.5}{123.5} = 0.78 \qquad (5.16)$$

It is usual to quote the value of $R^2$ rather than $R$ as it has the interpretation of being the fraction of total variance explained by the regression. Our calculation tells us that **78% of the variability of the girls' weights is explained by their dependence on their height and age according to the regression model constructed.**

The significance of this fraction can be judged by testing the variance it explains against the residual variance it does not explain. An analysis of variance table can be set up in the same form as Table (4.5) with the numerator of (5.15) as the regression sum of squares. The $F$ ratio introduced there is, in the multivariate case,

$$F = \frac{R^2(n-p-1)}{(1-R^2)p} \qquad (5.17)$$

where $n$ is the sample size and $p$ is the number of independent variables. It has $p$ and $(n-p-1)$ degrees of freedom. For our example,

$$F = \frac{0.78(8-2-1)}{(1-0.78)2} = 8.9 \qquad (5.18)$$

This has 2 and 5 degrees of freedom and from the tables on page 191 is significant at the 5% level. **Hence the regression model accounts for a significantly high proportion of the total variability.**

When there is only one independent variable the multiple correlation coefficient is the same as the correlation coefficient. The $F$ ratio is then the same as that in Chapter 4. The multiple correlation coefficient, however, does not test the significance of any one individual regression variable but the overall effect of the entire regression equation in describing the data. Sometimes it is informative to examine the residuals or errors which are the observed $y$ values minus the predicted ones. These can be plotted as a graph against $y$ or any of the independent variables. It is also possible to test the coefficient of each variable individually and this is shown in the next example. Many computer packages perform these tests automatically.

## Selection of variables in a multivariate regression

There are occasions when the sheer number of variables is daunting, as in the analysis of questionnaire data. Decisions must be taken as to which variables to include in a regression and clearly there is a need here for the clustering techniques discussed earlier in the chapter. The need is all the greater when the problem of multicollinearity described in the previous example is encountered.

Several methods for finding an optimal set of independent variables upon which to base a regression model have been proposed. The **forward stepwise** method is to begin with that variable having the highest correlation coefficient with the dependent variable. Regressions are then performed with each of the others acting as the second variable. The one giving the largest multiple regression coefficient is selected. Subsequent members of the set are chosen in this way until some criterion is satisfied—for instance that the multiple regression coefficient reaches some suitably high value. Unfortunately the mere inclusion of more variables in a regression raises the value of the coefficient irrespective of whether the additional ones contain relevant information for prediction purposes or not. Some computer packages generate an 'adjusted for degrees of freedom' multiple correlation coefficient to take this phenomenon into account.

The **backwards stepwise** method is to start with a regression on all the variables and consider the reduction in the multiple correlation coefficient caused by eliminating each one in turn. The one giving the smallest reduction is rejected from the set. Again the process is repeated until a sufficiently small set is obtained or some other criterion is satisfied.

A third, more powerful, method is presented in the next example. The **method of all possible regressions** is to implement what its name suggests and perform all possible regressions. For $p$ independent variables this entails $2^p-1$ hefty calculations but provided $p$ is less than about 14 a computer can supply solutions in a reasonable amount of time. This method is preferable to stepwise methods as the latter can miss the optimal set of variables completely and in any event often disagree in their estimates of what it is.

### The bolts example

A bolt manufacturing machine has three settings which affect the length of the bolts it produces. There is a centraliser, a cam angle and a pitch adjuster. Twenty bolts were made on the machine with the controls at various settings:

| Bolt length (mm) $y$ | Centraliser (mm) $x_1$ | Cam angle (°) $x_2$ | Pitch adjuster (mm) $x_3$ |
|---|---|---|---|
| 150 | 98 | 50 | 93.0 |
| 131 | 110 | 73 | 69.1 |
| 99 | 61 | 59 | 81.3 |
| 121 | 98 | 40 | 40.6 |
| 62 | 41 | 59 | 49.1 |
| 56 | 56 | 51 | 32.7 |
| 57 | 71 | 42 | 53.9 |
| 83 | 72 | 34 | 49.1 |
| 138 | 99 | 53 | 34.6 |
| 166 | 124 | 55 | 86.6 |
| 50 | 92 | 67 | 51.1 |
| 56 | 47 | 44 | 35.1 |
| 66 | 49 | 41 | 50.0 |
| 103 | 59 | 61 | 83.7 |
| 156 | 97 | 59 | 86.6 |
| 138 | 68 | 37 | 54.6 |
| 34 | 36 | 65 | 59.1 |
| 114 | 124 | 73 | 63.7 |
| 27 | 65 | 67 | 44.2 |
| 117 | 41 | 41 | 61.9 |

We examine the results of performing all possible regressions to see which set of the independent variables acts as the best predictor of bolt length.

*Solution* The raw data has been quoted in this example so that the reader can reproduce the calculations by computer or by hand if desired. The squares of the multiple correlation coefficients for each regression are displayed in Table 5.6 together with their $F$ ratios and degrees of freedom.

**Table 5.6** All possible regressions for the Bolts Example

| Variables included in regression | | | Multiple correlation coefficient squared | $F$ ratio | Degrees of freedom | | Significant at 5% |
|---|---|---|---|---|---|---|---|
| Centraliser $x_1$ | Cam angle $x_2$ | Pitch adjuster $x_3$ | | | | | |
| √ | — | — | 0.425 | 13.3 | 1 | 18 | yes |
| — | √ | — | 0.008 | 0.1 | 1 | 18 | no |
| — | — | √ | 0.328 | 8.8 | 1 | 18 | yes |
| √ | √ | — | 0.516 | 9.1 | 2 | 17 | yes |
| √ | — | √ | 0.566 | 11.1 | 2 | 17 | yes |
| — | √ | √ | 0.398 | 5.6 | 2 | 17 | yes |
| √ | √ | √ | 0.718 | 13.6 | 3 | 16 | yes |

We see immediately that the multiple correlation coefficient for the regression with all three independent variables is the highest. The second highest is for the regression omitting the cam angle and the significance of that omission can be tested. The $F$ ratio

$$F = \frac{(R^2 - R^2_{\text{without}\,x})(n-p-1)}{(1-R^2)} \tag{5.19}$$

with 1 and $(n-p-1)$ degrees of freedom tests whether the deletion of the variable $x$ from a regression on $n$ cases with $p$ variables gives a significant reduction in $R^2$. The square root of this ratio is a $t$ statistic and it is often quoted for each variable in the output from a computer package. Just as in the two-variable situation, we are really testing whether the slope of the regression plane due to $x$ is significantly different from zero or not. Applying the test to the omission of the cam angle from the set of all three variables:

$$F = \frac{(0.718 - 0.566)(20-3-1)}{(1-0.718)} = 8.6 \tag{5.20}$$

The $F$ table shows that the critical value at 5% for 1 and 16 degrees of freedom is 4.49 and so there is a significant reduction in the multiple correlation coefficient when the cam angle is left out of the calculations. **We conclude that all three variables should be included in the regression model**. Readers with access to a computer may wish to verify that the appropriate equation is

$$y = 38.0 + 0.96x_1 - 1.49x_2 + 1.12x_3$$

An $F$ ratio like (5.19) can be calculated to test the significance of a complete group of $m$ independent variables. The expression

$$F = \frac{(R^2 - R^2_{\text{without group}})(n-p-1)}{(1-R^2)m} \tag{5.21}$$

has $m$ and $(n-p-1)$ degrees of freedom.

## Summary

**1** Cases and variables can be clustered and possibly eliminated in order to reduce the complexity of the data. Case profiles may help with grouping cases by showing how their rankings on different variables compare with each other. If a distance function is defined between cases or variables then a minimal spanning tree can be drawn. Euclidean distance is often used for this purpose and for variables it is related to the correlation coefficient. The correlation matrix itself shows how variables are related. A dendrogram or tree diagram is a way of representing the clusters and the distances at which they form.

**2** The first principal component of a set of variables is the linear combination of them which accounts for the most variance in the sample. The second component accounts for the most variance once the effect of the first has been removed, and so on. They are usually calculated on a computer. Each one corresponds to an eigenvector of the correlation matrix, the coordinates of the vector being the multiples of the variables in the linear combination. The eigenvalue of the eigenvector expressed as a fraction of the total number of variables is the fraction of the total variability explained by that principal component.

**3** A multiple regression is an attempt to express one variable as a linear combination of predictor variables. Estimates of the coefficients are found by solving the normal equations which incorporate the sums of squares and products of the data values. Frequently the equations are ill-conditioned and need special care over accuracy in their solution. It may be necessary to exclude variables which are highly correlated with some of the others to avoid multicollinearity.

**4** The multiple correlation coefficient and its $F$ statistic are analogous with the two-variable coefficient and its $F$ statistic. The square of the multiple correlation coefficient is the fraction of the total variability explained by the model. It describes the overall regression and not the effect of any one variable.

**5** There are several methods for selecting the best set of predictor variables for a regression equation. Stepwise techniques introduce or eliminate variables one by one but, whenever feasible, performing all possible regressions is preferable. The contribution of an individual variable to a model can be tested for significance by an $F$ ratio based on the multiple correlation coefficient obtained by including it and the one obtained by excluding it. This test can be extended to groups of variables.

## Further reading

A special section on multivariate techniques has been included in the Bibliography as many elementary texts omit the subject completely. Reference (37) has plenty of worked examples on cluster analysis together with FORTRAN program listings for many useful algorithms, while (41) is about the grouping of plants and animals into species (taxonomy). The other books in the multivariate techniques section cover principal components and multiple regression, (35) being the authoritative text on the latter topic. As we saw in Chapter 4, variables can be transformed to linearise their relationship and often the new variables are analysed alongside the original ones. The methods of multiple regression are used to select those functions, usually powers, trigonometric ratios, exponentials or logarithms, which have the most significance in the model. References (9), (12) and (16) as well as those on regression in the 'Multivariate Techniques' section deal with such curvilinear regression.

Books like (39) describe methods not covered in this chapter. Factor analysis is an attempt to express variables as combinations, of factors, each one having a 'loading' on each of the factors. The assumption behind this approach is that there is a relatively small set of 'causes' which explain the observed correlations between the variables. A canonical correlation analysis is an extension of a regression in which there are several dependent variables. Combinations of them are found which have the highest correlation with the independent variables and so can be predicted accurately. These references also give a more detailed account of discriminant analysis than has been done here.

Specialised tests on applications to other disciplines often contain good treatments of problems like multicollinearity which are particularly relevant to those disciplines. The subject of econometrics deals almost exclusively with the construction of regression models to describe national and international economies and corporate behaviour. Reference (59) tackles multicollinearity, as many economic indicators are highly correlated with each other and cause the normal equations to degenerate. The phenomenon of heteroscedasticity, whereby different variables have different variances and are therefore not directly comparable with each other, is also discussed there. Another specialised area with its own set of difficulties is the statistical interpretation of questionnaire data. This is covered in more depth in Chapter 9, while Chapter 11 has a section on matrix and computer methods for implementing multivariate techniques.

## Exercises

**1** Draw case profiles for the following painters and classify them:

| Painter | Country | Date of birth | Type of work |
|---|---|---|---|
| Botticelli | Italy | 1445 | religious |
| Boucher | France | 1703 | portraits |
| Canaletto | Italy | 1697 | landscapes |
| Cézanne | France | 1839 | landscapes and still life |
| Dégas | France | 1834 | figures |
| Gainsborough | Great Britain | 1727 | landscapes and portraits |
| Gauguin | France | 1848 | figures |
| Giotto | Italy | 1267 | religious |
| Hogarth | Great Britain | 1697 | figures |
| Leonardo da Vinci | Italy | 1452 | religious and portraits |
| Manet | France | 1823 | landscapes and figures |
| Matisse | France | 1869 | figures |
| Michelangelo | Italy | 1475 | religious |
| Raphael | Italy | 1483 | religious |
| Rembrandt | Netherlands | 1606 | portraits |
| Renoir | France | 1841 | figures |
| Stubbs | Great Britain | 1724 | landscapes |
| Titian | Italy | 1487 | landscapes and figures |
| Van Gogh | Netherlands | 1853 | landscapes and figures |
| Vermeer | Netherlands | 1632 | figures |

**2** Cluster the patients in the Psychiatry Example.
**3** Using the data given in Question 5 of the Exercises to Chapter 2, derive a regression model of sales on year and advertising expenditure for the previous year. This is called a **lagged regression**. Test the model for significance.
**4** A glue has two active ingredients and differing quantities of them yielded the following strengths:

| Strength (N/mm$^2$) $y$ | Sticktite (kg) $x_1$ | Lockon (kg) $x_2$ |
|---|---|---|
| 29.4 | 3.4 | 2.1 |
| 33.3 | 4.3 | 2.5 |
| 26.4 | 4.0 | 2.6 |
| 28.1 | 3.9 | 2.4 |
| 24.2 | 3.2 | 2.7 |
| 27.5 | 3.5 | 1.9 |
| 32.6 | 4.4 | 2.3 |

Perform a regression of the strength on the quantities of the ingredients and analyse the significance of your model. Determine whether either of the predictors can be omitted by calculating the individual regressions of $y$ on $x_1$ and $x_2$ and using the 'all possible regressions' method.

# 6 Two samples

Previous chapters have been concerned with the analysis of a single sample of items drawn from a single homogeneous population. When an investigation has two different populations of interest or contains a factor which can take one of two levels for each subject, then two separate samples of data values will emerge. The presence of a two-level factor can be thought of as defining two distinct populations, one for each level.

The data from the two samples may be in the form of matching pairs. For instance in a study of the reasons for divorce one sample might be the husbands and the other the wives. The behaviour of each subject will then relate very strongly to the behaviour of the corresponding subject in the other sample. This effectively reduces the data to being a single sample; this would consist of 'couples' in the divorce study, and is dealt with in the section on Paired Data in Chapter 4. We assume here that while the samples are relevant to each other in information content they are statistically independent in their behaviour.

The objective in analysing two samples is usually to determine whether the two populations from which they were drawn are significantly different in some way. One of the samples might be a control group of subjects receiving a treatment known to be ineffective, against which a sample of test treatments can be assessed. Although it is preferable to match the subjects in the two samples for this purpose, there are situations when it is not possible to do so. The first worked example is a case in point and has data measured on a nominal or attribute scale. The frequencies of occurrence of the attributes, or equivalently their proportional occurrence, are compared across the two samples.

In the second example the measurement level is ordinal and a ranking method is employed. The third example is about the comparison of means and variances for two samples of metric data using $t$ and $F$ tests. There follows an example on the analysis of two samples with two variables each and the chapter finishes with a consideration of population size estimation from 'capture/recapture' data.

## Testing proportions

If the level of measurement is nominal, then the frequencies with which each attribute occurs in the two samples can be compared.

### The vaccine example

A group of 423 people received a vaccine against a certain disease. Their fortunes are tabulated together with those of another sample of 390 people who were not vaccinated:

| Number of people | Free of disease | Mildly affected | Strongly affected | Total |
|---|---|---|---|---|
| Not vaccinated | 282 | 73 | 35 | 390 |
| Vaccinated | 338 | 65 | 20 | 423 |

*Question 1* Is the vaccine effective in reducing the severity of the disease?

*Solution* Contingency tables were introduced in Chapter 4, where two variables measured on a single sample were analysed. Here the rows represent the two samples and testing for an association between row and column classifications is equivalent to comparing the samples for differences in the way the attributes are distributed. Table 6.1 is the result of applying the techniques of Chapter 4 for the null hypothesis of no association.

**Table 6.1** Calculation of expected frequencies for Question 1 of the Vaccine Example

| | Free of disease | Mildly affected | Strongly affected | Total |
|---|---|---|---|---|
| Not vaccinated | 282<br>(297.4) | 73<br>(66.2) | 35<br>(26.4) | 390 |
| Vaccinated | 338<br>(322.6) | 65<br>(71.8) | 20<br>(28.6) | 423 |
| Total | 620 | 138 | 55 | 813 |

The $\chi^2$ statistic for this table is 8.26 and there are 2 degrees of freedom. The tables on page 188 show that this is significant at the $2\frac{1}{2}\%$ level and so **there is evidence that the severity of the disease is associated with the vaccine.**

The $\chi^2$ test of association gives a positive result if the vaccine either raises or lowers a person's resistance to the illness. The term 'association' means any connection, not necessarily the one we are seeking. It is clear from an examination of the data in this example that the association is of the type we want. The next question illustrates the one-tail or two-tail testing of two proportions using the normal distribution.

*Question 2* Is the proportion of people not contracting the disease in the vaccinated sample significantly bigger than that in the unvaccinated one?

*Solution* Merging the two categories 'Mildly affected' and 'Strongly affected' to form a single category 'Affected by disease' gives Table 6.2.

**Table 6.2** Merged frequencies for Question 2 of the Vaccine Example

| | Free of disease | Affected by disease | Total |
|---|---|---|---|
| Not vaccinated | 282 | 108 | 390 |
| Vaccinated | 338 | 85 | 423 |

The vaccinated sample 'success rate' is (338/423), which is 0.7991, and the unvaccinated sample proportion is (282/390) which is 0.7231. We wish to know whether the percentage 79.9 measured on a sample of size 423 is significantly bigger than a percentage of 72.3 measured on a sample of size 390.

We saw in the Prime Minister Example of Chapter 3 that a proportion measured on a large sample can be assumed to be normally distributed. As the sum or difference of independent normally distributed variables is itself normally distributed (see the

Rice Example of Chapter 3) we can test the change in the proportions using that distribution. The correction for continuity when applying the normal approximation consists in this case of subtracting 0.5 from the numerator of the larger proportion and adding 0.5 to the numerator of the smaller one. We obtain (337.5/423) and (282.5/390) in this example, which are equal to 0.7979 and 0.7244 respectively, and we shall call them $p_1$ and $p_2$. The relevant $z$ score for samples of size $n_1$ and $n_2$ to test the null hypothesis that the two population proportions are the same is:

$$z = \frac{p_1 - p_2}{\sqrt{\left[g(1-g)\left(\frac{1}{n_1}+\frac{1}{n_2}\right)\right]}} \tag{6.1}$$

The quantity $g$ in this formula is an estimate of the common proportion. The best estimate for this based on the null hypothesis is obtained by pooling the data from the two samples:

$$g = \frac{338+282}{423+390} = \frac{620}{813} = 0.7626 \tag{6.2}$$

The $z$ score is therefore

$$z = \frac{0.7979 - 0.7244}{\sqrt{[0.7626(1-0.7626)(\frac{1}{423}+\frac{1}{390})]}} = 2.46 \tag{6.3}$$

We have calculated this $z$ score to measure the *increase* in a proportion from one sample to another and we therefore perform a one-tail test. From the standard normal tables on page 183 we see that a $z$ value as large or larger than 2.46 has a probability of only 0.00695. Thus we can reject the null hypothesis that the proportions are the same at the 1% level of significance. **The vaccine does indeed prevent the occurrence of the disease.**

The relationship between the normal distribution and the $\chi^2$ with one degree of freedom was discussed in the Hand Cream Example of Chapter 3. The $z$ score approach allows a one-tail test to be performed and also confidence intervals to be calculated for the difference in proportions if required. The denominator of (6.1) gives the standard deviation for such a calculation which would utilise equation (3.2), as in the Rice Example.

## The Mann–Whitney test

When the categories being measured are on an ordinal scale, ranking methods can be used. The rank test developed by Wilcoxon for a single sample and considered in Chapter 4 was applied by him to two samples. In view of additional work carried out by them, however, the procedure has since become known as the Mann–Whitney test.

### The house prices example

A selection of houses sold in England and Wales had the following prices:

| | Selling price (£ thousands) | | | | | | | | |
|---|---|---|---|---|---|---|---|---|---|
| England | 54 | 72 | 63 | 60 | 61 | 58 | 84 | | |
| Wales | 29 | 38 | 65 | 59 | 71 | 66 | 61 | 59 | 63 |

The estate agent who collected the data feels that the houses could have been sold for a few thousand pounds more or less than the actual figures. The prices give only a rank ordering to the properties and should not be treated as accurate metric measurements. We investigate whether the Welsh house prices are significantly lower than the English ones.

*Solution* Ranking all the data enables us to see to what extent the two samples become mixed together (Table 6.3).

**Table 6.3** Ranked prices in the House Prices Example

| Rank | 1 | 2 | 3 | 4 | 5.5 | 5.5 | 7 | 8.5 | 8.5 | 10.5 | 10.5 | 12 | 13 | 14 | 15 | 16 |
|---|---|---|---|---|---|---|---|---|---|---|---|---|---|---|---|---|
| Price | 29 | 38 | 54 | 58 | 59 | 59 | 60 | 61 | 61 | 63 | 63 | 65 | 66 | 71 | 72 | 84 |

Tied values are awarded the average of the ranks they would have otherwise occupied. The prices of the Welsh houses are underlined in Table 6.3 and we want to know if they are towards the beginning of the sequence or not. The total of their ranks is

$$T = 1+2+5.5+5.5+8.5+10.5+12+13+14 = 72 \tag{6.4}$$

Now an unusually small value of $T$ would indicate that the Welsh prices tend to be at the lower end of the rank order. Provided both samples have more than 5 data values each the statistic $T$ is approximately normally distributed and so we can test its significance. The $z$ score for a $T$ value based on a sample of size $n$ is

$$z = \frac{T-(n+m+1)n/2}{\sqrt{[nm(n+m+1)/12]}} \tag{6.5}$$

where $m$ is the size of the other sample. If $z$ is less than 2 then a continuity correction consisting of subtracting 0.5 from the absolute size of the numerator should be applied.

For our data we find that

$$z = \frac{72-(9+7+1)9/2+0.5}{\sqrt{[9\times 7\times(9+7+1)/12]}} = -0.423 \tag{6.6}$$

From the standard normal tables on page 183 it follows that the probability of $T$ being less than or equal to 72 is 0.33724. To be significant, however, the value 72 must lie in a low-probability region and so **we have no evidence that the price of Welsh houses is significantly lower than that of English ones.**

For samples which are too small for the normal distribution to apply, a direct enumeration of rank sums can be used to test the significance of the $T$ statistic. For instance suppose a sample of 3 items gave a rank sum of 8 and the other sample contained 4 items. There are 7 data values altogether and we need to evaluate the probability that 3 of them chosen at random have ranks adding up to 8 or less. There are just 4 ways that 3 ranks from 1 to 7 can add up to 8 or less—namely 123, 124, 125 and 134. The total number of ways 3 ranks can be chosen from 7 is $(7!/4!3!)$ which is 35, and so the probability that 3 ranks sum to 8 or less by chance is $\frac{4}{35}$. We can therefore attach a probability to the occurrence of a $T$ statistic as small or smaller than the one we observed and decide on its significance. In general there are $(n+m)!/n!m!$ ways of choosing $n$ ranks from $(n+m)$.

## The two-sample *t* test

For metric data it is often possible to make the normality assumption of classical sampling theory. The $F$ and $t$ distributions then provide tests of the similarity between the two populations represented by the samples.

### The cats example

For 6 months 12 cats were given a tinned food, 'Fatpuss', and the gains in their weight over that period were recorded. A second sample of 14 cats was fed on a rival brand, 'Catto', and similar measurements made. The results were:

| | |
|---|---|
| Fatpuss sample weight gains (g) | 162, 189, 183, 194, 201, 188, 182, 187, 193, 209, 214, 191 |
| sample size=12; sum of $(x)$=2293; sum of $(x^2)$=440 135<br>sample mean=191.083; sample variance=180.083 | |
| Catto sample weight gains (g) | 191, 188, 222, 209, 221, 234, 206, 198, 180, 186, 177, 201, 223, 220 |
| sample size=14; sum of $(x)$=2856; sum of $(x^2)$=586 882<br>sample mean=204.000; sample variance=327.538 | |

*Question 1* Do the variances of the samples indicate that the two populations from which they came have the same variability?

*Solution* A test for common variability should always be carried out before the means of two or more samples are compared. The outcome of the test affects the subsequent analysis in that either the variances can be pooled or they must be treated as essentially different estimates of different parameters. Apart from this consideration it is important to the researcher to know whether the populations on which the treatments are being tested are alike in the way they are hosting those treatments.

In the analysis of variance for regressions of Chapters 4 and 5 the $F$ ratio was used to decide whether two variances differ significantly or not. Putting the larger variance on the top of the ratio we obtain

$$F=\frac{327.538}{180.083}=1.82 \tag{6.7}$$

The number of degrees of freedom of the larger variance, $\nu_1$, is $(14-1)$, while that of the smaller variance, $\nu_2$, is $(12-1)$. Taking the nearest tabulated numbers of degrees of freedom to these in the table on page 191, we find the 5% critical value for an $F$ ratio to be 2.72. **There is thus no evidence to suggest that the variances are significantly different because our $F$ ratio is less than the critical value.**

*Question 2* Do the means of the samples indicate a significant difference between the weight gains due to the two foods?

*Solution* If the $F$ ratio tested above is not significantly large, then the variances are both estimates of a common 'weight gain' variability and a **pooled estimate** can be calculated:

$$s_p^2 = \frac{(n_1-1)s_1^2+(n_2-1)s_2^2}{(n_1+n_2-2)} \tag{6.8}$$

Here $n_1$, $s_1^2$, $n_2$ and $s_2^2$ have the obvious interpretation of being the size and variance of the two samples. The two-sample $t$ statistic measuring the difference between the sample means is

$$t = \frac{(\bar{x}_1-\bar{x}_2)-(\mu_1-\mu_2)}{s_p\sqrt{[(1/n_1)+(1/n_2)]}} \tag{6.9}$$

where $\mu_1$ and $\mu_2$ are the population means. This has $(n_1+n_2-2)$ degrees of freedom because the two means $\bar{x}_1$ and $\bar{x}_2$ have been calculated from the original $(n_1+n_2)$ data values.

If on the other hand the variances are significantly different, then the score

$$z = \frac{(\bar{x}_1-\bar{x}_2)-(\mu_1-\mu_2)}{\sqrt{[(s_1^2/n_1)+(s_2^2/n_2)]}} \tag{6.10}$$

can be used provided that the samples are both large. There are no reliable techniques for small samples with dissimilar variances.

It is preferable to be able to pool the variances according to (6.8) as it effectively increases the sample size on which subsequent tests are based. In the present example, pooling the variances gives

$$s_p^2 = \frac{(12-1)180.083+(14-1)327.538}{(12+14-2)} = 259.954 \tag{6.11}$$

We can now test the null hypothesis that $\mu_1$ is equal to $\mu_2$ and there is no difference between the population weight gains due to the cat foods: equation (6.9) becomes

$$t = \frac{(191.083-204.000)-(0)}{\sqrt{259.954}\sqrt{(\frac{1}{12}+\frac{1}{14})}} = -2.036 \tag{6.12}$$

The critical two-tail value for $t$ at the 5% level with 24 degrees of freedom is 2.064 from the table on page 192. **Hence we cannot reject the null hypothesis that the means are the same, the cat foods do not appear to be different from each other.**

The $t$ and $z$ statistics defined by (6.9) and (6.10) can be used to perform one-tail as well as two-tail tests. For instance a large positive $t$ value is evidence in favour of the alternative hypothesis that $\mu_1$ is greater than $\mu_2$. Furthermore the techniques involving inequalities in the Crisps and Train Journey Examples of Chapter 3 can be applied to derive confidence intervals for the difference between the two population means.

## The analysis of covariance

The method of linear regression enables us to express the relationship between the variables in a single sample in a mathematical form. Sections of Chapters 4 and 5 are devoted to this important branch of statistics. For two or more samples, the results of the individual regressions can be compared with each other. Analysis of covariance is sometimes referred to as **ANCOVA**.

**The vitamin example**

A sample of men and a sample of women were given various doses of a certain vitamin. Their responses, measured in milligrams of carbolic acid, were:

| | Women | | Men | |
|---|---|---|---|---|
| | Dose (ml) $x_1$ | Response (mg) $y_1$ | Dose (ml) $x_2$ | Response (mg) $y_2$ |
| | 3.4 | 17.9 | 2.2 | 15.7 |
| | 3.7 | 14.7 | 2.5 | 11.9 |
| | 3.9 | 20.4 | 2.6 | 16.3 |
| | 4.1 | 17.0 | 2.6 | 20.1 |
| | 4.4 | 15.5 | 2.9 | 17.0 |
| | 4.8 | 16.9 | 3.1 | 21.4 |
| | 5.2 | 21.5 | 3.5 | 18.3 |
| | 5.4 | 20.7 | 3.7 | 22.2 |
| | 5.8 | 21.4 | 3.9 | 19.3 |
| | 5.9 | 24.5 | 4.0 | 20.6 |
| | 5.9 | 20.4 | | |
| | 6.2 | 23.5 | | |
| | 6.4 | 24.1 | | |
| Sum | 65.1 | 258.5 | 31.0 | 182.8 |
| Mean | 5.01 | 19.88 | 3.10 | 18.28 |
| Sum of squares | 338.53 | 5265.69 | 99.78 | 3429.34 |
| Sum of $xy$ | 1326.38 | | 578.43 | |
| Sample size | 13 | | 10 | |

In essence we want to compare the responses between men and women after they have been adjusted for the effect of the dosage level. Dose is called a **concomitant** variable in this context as it accompanies the dependent variable and contributes to its value. There may be more than one concomitant variable, in which case the analysis of covariance relates to multiple regressions performed on the individual samples.

*Question 1* Plot scatter diagrams for the two samples on the same axes and calculate separate regression lines for them.

*Solution* Figure 6.1 shows the points and lines required. The reader is referred to the Cholesterol Example of Chapter 4, in particular Table 4.3, for an explanation of the methodology. The corrected sums needed are given in Table 6.4 and can be verified for practice from the information contained in the data table. **The equations are $y_1 = 7.10 + 2.55x_1$ and $y_2 = 8.39 + 3.19x_2$.**

*Question 2* Do the responses, after they have been adjusted for dose, have significantly different variability between the two samples?

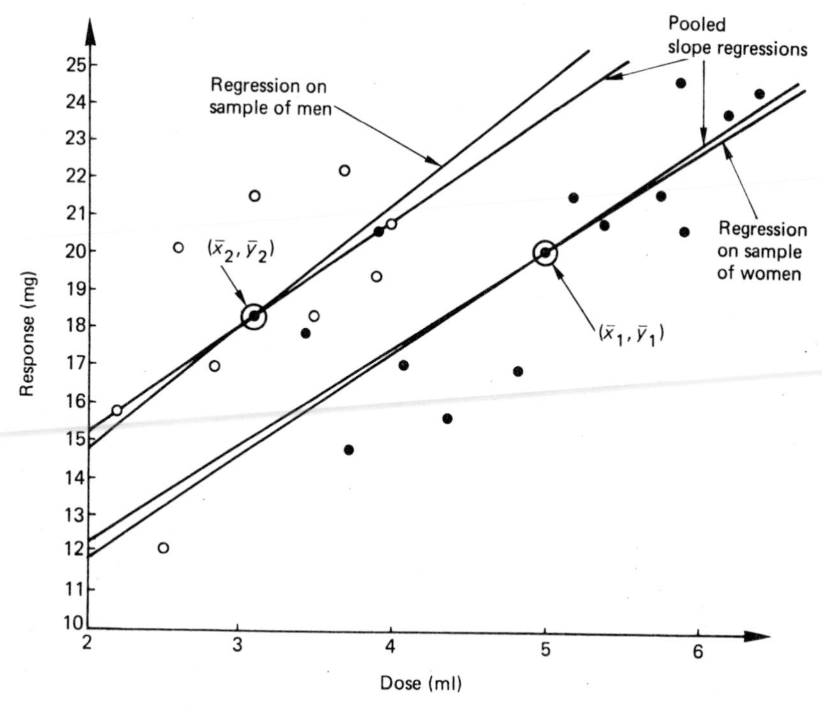


**Fig. 6.1** Regressions for the Vitamin Example

*Solution* As for a two-sample $t$ test, it is important to compare variances before any pooling of the data is considered. If the two samples are so dissimilar that they indicate populations which are fundamentally different, then the analysis would not be carried any further. The populations are distinct in such situations and nothing more can be said.

The residual unexplained variance remaining in the dependent variable after a regression model has been fitted was defined in the Heater Example of Chapter 4. For multiple regressions we saw in Chapter 5 that the residual sum of squares is $(1-R^2)CS_{yy}$. If there is just one independent variable, as here, the multiple correlation coefficient $R$ is the same as the ordinary correlation coefficient $r$ and the two formulae give identical results. Table 6.4 shows the corrected sums for the individual regressions and the consequent residual sums of squares. Each of the first two rows is an analysis of variance like Tables 4.5 and 4.6. The $F$ ratios for the individual regressions in column 10 are both significantly large: the first has 1 and 11 degrees of freedom, the second 1 and 8. This implies that the responses in each sample can be satisfactorily explained by linear models.

The variability of the responses after they have been adjusted for the concomitant variable, dose, is measured by the residual mean square. We can compare them for the two samples by an $F$ ratio of the larger divided by the smaller:

$$F_{\text{residual variances between individual regressions}} = \frac{6.28}{4.03} = 1.6 \qquad (6.13)$$

This has 8 and 11 degrees of freedom and from tables is not significant. Although we have not yet tested the difference in the average responses or in the ways in which response depends on dose, **we can say at this stage that the background variability in**

**Table 6.4** Analysis of covariance table for the Vitamin Example

| Column | (1) | (2) | (3) | (4) | (5) | (6) | (7) | (8) | (9) | (10) |
|---|---|---|---|---|---|---|---|---|---|---|
| Source | $CS_{xx}$ | $CS_{xy}$ | $CS_{yy}$ | Degrees of freedom $n-1$ | Slope $b$ (2) ÷ (1) | Regression sum of squares $(2)^2 \div (1)$ | Residual sum of squares (3) − (6) | Residual degrees of freedom $n-2$ | Residual mean square (7) ÷ (8) | $F$ ratio (6) ÷ (9) |
| Regression on women's sample | 12.529 | 31.892 | 125.517 | 12 | 2.55 | 81.180 | 44.337 | 11 | 4.03 | 20.14 |
| Regression on men's sample | 3.680 | 11.750 | 87.756 | 9 | 3.19 | 37.517 | 50.239 | 8 | 6.28 | 5.97 |
| | Total residual sum of squares and degrees of freedom | | | | | | 94.576 | 19 | 4.98 | |
| Between individual slopes and pooled slope (by subtraction) | | | | | | | 1.193 | 1 | 1.19 | |
| Regressions with common pooled slope | 16.209 | 43.642 | 213.273 | 21 | 2.69 | 117.504 | 95.769 | 20 | 4.79 | |
| Between pooled slope and overall regression (by subtraction) | | | | | | | 31.074 | 1 | 31.07 | |
| Overall regression on all data | 36.779 | 60.943 | 227.826 | 22 | 1.66 | 100.983 | 126.843 | 21 | 6.04 | 16.72 |

**both populations can be assumed to be the same.** The residuals of multiple regressions can be treated in this manner whilst if there are more than two samples the $F$ test can be applied to them in pairs.

*Question 3* Is there a significant difference between the slopes of the two individual regression lines?

*Solution* Once we are convinced that the samples are comparable with each other the regression lines themselves can be examined. The full meaning and beauty of Table 6.4 can now be described. The first two rows are the separate sample regressions. The last row is a regression over all the $x$ and $y$ values treated as a single sample. This represents a complete amalgamation of the data and assumes that there is a common slope and a common intercept to the individual lines. A covariance analysis helps us to discover how far down the table we can go in pooling slopes and intercepts.

Regression lines for each sample utilising a common slope value are calculated in the fifth line of the table. The corrected sums and degrees of freedom are the totals for the individual regressions. A slope estimate and residual sum of squares are derived in the usual way. The excess of this residual sum of squares over and above the total of the individual regressions is due to the assumption of a common slope. In order to assess the difference between the original separate slopes and the pooled slope we can calculate the $F$ ratio:

$$F_{\text{between individual slopes and pooled slope}} = \frac{1.19}{4.98} = 0.24 \tag{6.14}$$

This has 1 and 19 degrees of freedom but even without tables we can see it is not significant as it is less than 1. The increase in residual variability introduced by the adoption of common slope models is not significantly large when compared with the existing combined residual. **There is thus no significant difference between the slopes, and a common pooled estimate is acceptable.**

*Question 4* Having decided that common slope models are feasible descriptors of the samples, plot them on the scatter diagram and determine whether their intercepts are significantly different.

*Solution* The common slope estimate is 2.69 from Table 6.4 and so, using the expression $(\bar{y} - b\bar{x})$ from Table 4.3 for the intercept, the lines are:

$$\begin{aligned} y_1 &= \bar{y}_1 - b\bar{x}_1 + bx_1 = 19.88 - (2.69)(5.01) + 2.69x_1 = 6.40 + 2.69x_1 \\ y_2 &= \bar{y}_2 - b\bar{x}_2 + bx_2 = 18.28 - (2.69)(3.10) + 2.69x_2 = 9.94 + 2.69x_2 \end{aligned} \tag{6.15}$$

These lines are drawn on the scatter diagram of Fig. 6.1.

As far as the vitamin being investigated is concerned, these models have been derived on the assumption of a common response rate per unit dose but acknowledging that the zero dose response, or that of any other standard level, may be different between men and women. Parallel line models are often encountered in the biological assay, or **bioassay**, of drugs when the relative potency of several samples are to be compared. The logarithm of the dose is frequently used for this purpose rather than the dose itself.

We are now half-way down Table 6.4 and have tested and accepted the pooled regressions of its fifth row of working. The intercepts of these common slope models can be examined by comparing them with an overall regression which has not only a single slope but also a single intercept. The calculation of such a line forms the bottom row of Table 6.4 and the increase in residual variability is obtained by subtraction in the previous row. The $F$ ratio to test this difference is:

$$F_{\text{between different intercepts and common intercept}} = \frac{31.07}{4.79} = 6.5 \tag{6.16}$$

This has 1 and 20 degrees of freedom and is significant at the 5% level. In other words, the fitting of an overall regression line produces an unacceptably large increase in the residual sum of squares. **The intercepts of the pooled slope models are significantly different from each other.** Notice that although the overall regression has a highly significant $F$ ratio it does not form as good a descriptive model of the data as the two lines with a common slope which we have identified.

## Population size estimation

Ecologists and zoologists need to be able to estimate the size of animal and plant populations from data collected at different locations or at different times. The observed numbers might be treated as a time series and regression analysis used to predict future values. More sophisticated forecasting techniques borrowed from operational research can help deal with seasonality and long-term cyclical behaviour. The problem of estimating the total population size from the numbers observed, however, is essentially a probabilistic one. A popular approach for animal species is to mark a sample, release them and then measure how many marked specimens there are in a second sample caught at a later date. This is the **mark, release and recapture** or **capture–recapture** method, and it is illustrated in the next example.

### The squirrels example

A biologist captures 24 grey squirrels from a wood and puts ink marks on their backs. One week later a second sample of 19 grey squirrels contains 5 with ink marks. Estimate the size of the grey squirrel population in the wood.

*Solution* Elsewhere in this book population parameters like means, variances and regression-line slopes are described in terms of point and interval estimates derived from a sample. In classical sampling theory the normality assumption has the effect of predetermining the probability distributions of sample statistics like $t$, $\chi^2$ and $F$. As these distributions are complicated it is inappropriate in a practical text like this one to delve into the precise nature of the estimation process using them. The present example, however, provides an opportunity to examine estimation procedures more closely with a simpler distribution.

### 1 The method of moments

The average value of $x^r$ for a sample of $x$ values is the **$r$th sample moment**. The expectation of $x^r$ for the entire population is the **$r$th population moment**. The **method of moments** is to equate corresponding sample and population moments to obtain equations for and hence estimates of population parameters.

The first moments are the sample and population means. Equating these simply implies taking the sample average to be an estimate of the population average. Equating second moments is equivalent to using the sample variance or standard deviation as an estimate of the population value.

After the biologist has marked the first sample of squirrels there are 24 with ink marks out of a total of $N$, where $N$ is the population size. The probability of a squirrel chosen at random being marked is therefore $(24/N)$ and as 19 are chosen in the second sample we expect this fraction of 19 to be marked. Hence the theoretical expectation of the number marked is $(24/N)$ times 19. According to the question the sample value of this first moment is 5 and so we obtain the equation:

$$\frac{24}{N} \times 19 = 5 \tag{6.17}$$

Making $N$ the subject of this equation produces the estimate of 91.2, or more sensibly 91, for the number of squirrels in the wood.

**2 The method of maximum likelihood**

The philosophy of this method is to find the value of the population parameters which give the observed data as big a probability as possible. For continuous distributions the maximum of the probability density function is located and parameter values corresponding to that maximum taken as estimates.

The 19 animals caught as the second sample constitute 19 trials for which the probability of being marked is $(24/N)$ for each one. The binomial probability distribution describes such situations as in the Toaster Example of Chapter 3. Applying Equation (3.22) with $n$ and $p$ equal to 19 and $(24/N)$ respectively we find the probability of exactly 5 successes is:

$$P(5 \text{ successes out of 19 trials}) = \frac{19!}{14!5!}(24/N)^5(1-(24/N))^{19-5} \tag{6.18}$$

This is called the **likelihood** of the sample and for a continuous random variable would be a probability density function. The **method of maximum likelihood** is to choose the value of $N$ which makes this as large as possible to be the estimate.

Readers familiar with calculus might care to differentiate (6.18), or more easily its natural logarithm, and by equating the derivative to zero find the required $N$ value. It turns out to be exactly the same as above, namely 91.2. **Although we have used two different methods, the estimates of the size of the population of squirrels are both 91.**

The two methods do not always give identical results and the maximum likelihood technique is to be preferred on theoretical grounds. Fortunately for users of practical statistics the estimation of parameters from first principles is not often necessary and standard procedures for means and variances can be employed.

## Summary

**1** Data is collected as two samples whenever there are two distinct populations or there is a factor with two levels. The main aim in the analysis is to determine whether the samples are sufficiently similar for them to be amalgamated in some way. The case of matched data was considered in Chapter 4.

**2** Two samples of attribute data can be compared in a contingency table. The difference in the proportional occurrence of an attribute between two samples is measured either by a $\chi^2$ statistic or a $z$ score. The Mann–Whitney test applies to two samples of ordinal data values. They are all ranked together, and if the samples merge sufficiently well, then the populations from which they were drawn are indistinguishable from each other.

**3** When the normality assumption can be made about the two populations, a two-sample $t$ test detects a significant difference between the means of the samples. The $t$

statistic itself should be used only when the variances have been pooled as a result of their $F$ ratio being small. For large samples with different variances a $z$ score can be constructed to describe the relationship between the means.

**4** The analysis of covariance allows regression lines on two or more samples to be compared. It can be easily extended to cope with multiple regressions. Having verified that the residuals are homogeneous, the data is pooled to give a common slope estimate. The discrepancies between individual slopes and the common slope are tested and then the hypothesis of a common intercept examined. If this can be accepted, then the overall regression model is satisfactory; otherwise parallel lines describe the dependence of one variable on another for each sample separately. An analysis of covariance is really an analysis of variance, which we discuss in the next chapter, on the dependent variable after it has been adjusted for the effect of the independent concomitant variable.

**5** Capture–recapture experiments are often used to estimate animal population size. The animals in the first sample are marked and merely 'set the scene' for the second sampling operation. The number of marked specimens caught is then incorporated in an estimation procedure which can utilise the method of moments or the method of maximum likelihood.

## Further reading

Contingency tables and two sample $t$ tests are covered in most elementary statistics texts. The Mann–Whitney test is given a thorough treatment in (30) and (33), the latter taking a more rigorous approach than the one adopted here. The theory of the analysis of covariance can be found in (21), while (19) contains useful material. Methods of bioassay are given in (12), which contains more references on this topic. Population size estimation is the subject of (53), which is included in the Bibliography as most statistics books pay only passing attention to the problem.

## Exercises

**1** Two pesticides were applied to separate samples of elm fleas:

| | Number surviving | Number dying |
|---|---|---|
| Pesticide A | 34 | 128 |
| Pesticide B | 58 | 125 |

(i) Perform a $\chi^2$ test to decide whether the pesticides are significantly different from each other in their effect. (ii) Calculate an equivalent $z$ score and verify it is the square root of the $\chi^2$ statistic found in (i). Use it to determine whether pesticide $A$ is significantly better than pesticide $B$ in killing elm fleas.

**2** The times taken for a group of experienced rats to run a maze are to be compared with the times for a group of inexperienced ones:

| | |
|---|---|
| Experienced rats' times (seconds) | 121, 137, 130, 128, 132, 127, 129, 131, 135, 130, 126, 120, 118, 125. |
| Inexperienced rats' times (seconds) | 135, 142, 145, 156, 149, 134, 139, 126, 147, 152, 153, 145, 144. |

(i) Perform a Mann–Whitney test to determine whether the inexperienced rats' times are significantly greater than the experienced rats' times. (ii) Perform a two-sample $t$ test for the same null and alternative hypotheses as (i) and comment. Remember that 100 can be subtracted from every data value without affecting the variances or the relationship between the two averages.

**3** A chemical engineer is comparing two processes for manufacturing a new plastic. He measures the melting points of several specimens made with different amounts of nitric acid for each process:

| Process A | | Process B | |
|---|---|---|---|
| Percentage nitric acid ($x_1$) | Melting point °C ($y_1$) | Percentage nitric acid ($x_2$) | Melting point °C ($y_2$) |
| 12 | 8.4 | 19 | 13.5 |
| 15 | 6.8 | 25 | 18.7 |
| 17 | 10.5 | 29 | 17.4 |
| 18 | 9.8 | 32 | 19.9 |
| 23 | 13.4 | 37 | 17.8 |
| 27 | 12.2 | 38 | 24.4 |
| 29 | 20.0 | 39 | 26.0 |
| 30 | 20.4 | 47 | 22.4 |
| 32 | 25.2 | 52 | 30.7 |
| 33 | 16.9 | 60 | 26.5 |
| 45 | 16.2 | | |
| 64 | 24.6 | | |
| 73 | 26.3 | | |

Perform an analysis of covariance to determine whether the processes are significantly different from each other in the melting points they produce.

# 7 Many samples

There will be more than two samples of data values if the investigation covers more than two populations, has one factor with more than two levels or has several factors. It will be remembered from Chapter 1 that to be considered as coming from the same sample data values must be directly comparable amongst themselves in every respect. Each combination of factor levels is called a treatment and produces a separate sample. In this chapter, because of the possibly large numbers of factor level combinations, that is treatments, we shall encounter samples with just one data value in them. Whilst methods of analysing multi-sample data structures are dealt with here, the next chapter is about ways of designing appropriate experiments or data-collection exercises.

Samples of frequencies of attribute variables can be analysed in the usual manner as a contingency table. There are numerous examples of such tables in this book. For instance the Cars Example of Chapter 4 can be thought of as the analysis of samples of vehicles from each of the four manufacturers. Although the attribute 'manufacturer' was treated there as a random variable the analysis would be exactly the same if four samples with predetermined sizes had been selected and then classified by 'type of fault' only.

The occurrence of many samples of ordinal data is relatively less common and will be dealt with in the second example of this chapter. The first, third and fourth examples are concerned with metric data. Although not covered here, several samples of two or more metric variables can be examined by a simple extension of the analysis of covariance technique described in Chapter 6. Analysis of variance for more than two factors is dealt with in the next chapter.

## One-way analysis of variance

Samples which differ from each other because of the level of just one factor are said to form a one-way classification scheme. By analysing the variability across the classification we can extend the two-sample $t$ test of the last chapter to compare the sample means with each other.

### The alloys example

Four alloys are being assessed for their strength. The tensile breaking strains are measured for various specimens:

| Alloy | Tensile breaking strain (coded units) |
|---|---|
| Irongrip | 32, 37, 34, 33, 37 |
| Brassmeld | 30, 31, 40, 37, 38, 36 |
| Tuffstuff | 24, 31, 28, 32 |
| Coldsteel | 30, 29, 28, 33, 31, 32 |

Although it is preferable for theoretical reasons to have all the samples of the same size it is often the case that they are not.

*Question 1* Are the alloys' breaking strains significantly different from each other?

*Solution* The model assumed for the data values in a one-way analysis of variance is:

$$x = \mu + \alpha_{\text{treatment}} + \text{error} \tag{7.1}$$

The four quantities $\alpha_{\text{treatment}}$ in this example represent the deviations of the mean breaking strain of each alloy from the overall mean $\mu$. The terms in such a model can be considered as fixed parameters, giving a **fixed-effects model**, or as random variables, giving a **random-effects model**. An effect is fixed if the factor it relates to is within our control. Our model has fixed effects as we are testing specimens of the alloys especially prepared for the purpose. If we chose a piece of metal at random and then determined which alloy it was made of, then 'alloy' would be a random variable and the distribution of the various alloys amongst the population we are sampling from would affect the data values and even the sample sizes we obtain. All the models in this chapter have fixed effects.

In applying classical sampling theory to the analysis of variance it is necessary to make certain assumptions regarding the error term in (7.1). Errors are taken to be normally distributed with zero mean and a variance which is the same for every treatment. Furthermore the errors for different measurements must be uncorrelated with each other. These restrictions on the methodology are rarely tested for validity in solving practical problems.

The value of $\mu$ in the model can be estimated by averaging all the data. Each $\alpha$ is then approximated by the treatment mean minus this estimate. The model can therefore be expressed in terms of averages calculated from the data:

$$x = \bar{x} + (\bar{x}_{\text{treatment}} - \bar{x}) + \text{error} \tag{7.2}$$

The difference between each $x$ value and the overall mean can thus be written:

$$x - \bar{x} = (\bar{x}_{\text{treatment}} - \bar{x}) + \text{error} \tag{7.3}$$

It is the relative size of the two parts of the right-hand side of this equation which are assessed by an ANalysis Of VAriance or **ANOVA**. In order to understand the procedure it is necessary to do a small amount of algebra.

Squaring both sides of equation (7.3) and summing them for all the data values gives:

$$CS_{xx} = CS_{\text{treatments}} + \text{residual error} \tag{7.4}$$

The left-hand side, $CS_{xx}$, is recognised as the corrected sum of squares introduced in Chapter 2 as the numerator of a variance calculation. It appears also in the regression analysis work of Chapter 4 and, as there, denotes the total variability of the data values from a central average.

The right-hand side of (7.4) contains the sum of the treatment deviations squared, $CS_{\text{treatments}}$, plus the sum of the squared errors, called the **residual error**. The product terms obtained from squaring (7.3) add up to zero because the sum of the errors within each sample is zero, from equation (7.2). Terms in a model whose products sum to zero are **orthogonal** and enable us to evaluate the contribution of each part of the model to the overall sum of squares easily. We can determine in (7.4) how much of the total corrected sum of squares the differences between the treatment means and the grand mean explain. The size of the residual is a measure of how much variability in the data we are consigning to statistical or sampling error.

A convenient form for the **treatments sum of squares,** $CS_{\text{treatments}}$, is:

$$CS_{\text{treatments}} = \text{sum of}\left(\frac{T_i^2}{n_i}\right) - \frac{(\text{sum of } T_i)^2}{(\text{sum of } n_i)} \tag{7.5}$$

where $T_i$ and $n_i$ are the total of the $x$ values and the number of $x$ values for treatment $i$. The calculations for equation (7.4) using this expression on the data for the alloys are shown in Table 7.1. The results are usually set out in an analysis of variance table and this is illustrated in Table 7.2. The number of degrees of freedom for the treatments is 3 because there are 4 treatments and their total has been used in calculating the corrected sum of squares. This implies that the treatment means have 4 minus 1 degrees of freedom. The total number of degrees of freedom is 20 as there are 21 data values and their overall mean has been computed, thus reducing the information content by 1. The number of degrees of freedom for the residual sum of squares is found by subtraction. The column headed 'Mean square' in Table 7.2 is the sum of squares divided by the number of degrees of freedom while the $F$ ratio is the treatments mean square divided by the residual mean square. The complete table can be compared with Tables 4.5 and 4.6 which display the analysis of variance for a regression line.

**Table 7.1** Analysis of variance calculations for the Alloys Example

| Treatment (alloy) | Irongrip | Brassmeld | Tuffstuff | Coldsteel | |
|---|---|---|---|---|---|
| | 32<br>37<br>34<br>33<br>37 | 30<br>31<br>40<br>37<br>38<br>36 | 24<br>31<br>28<br>32 | 30<br>29<br>28<br>33<br>31<br>32 | |
| Treatment total, $T_i$ | 173 | 212 | 115 | 183 | sum of $T_i$=683 |
| Sample size, $n_i$ | 5 | 6 | 4 | 6 | sum of $n_i$=21 |
| Treatment mean, $\bar{x}_i$ | 34.60 | 35.33 | 28.75 | 30.50 | $\bar{x}=\frac{683}{21}=32.52$ |

Sum of $(x^2)=22\,521$; $CS_{xx}=22\,521-\dfrac{683^2}{21}=307.238$

$$CS_{\text{treatments}}=\frac{173^2}{5}+\frac{212^2}{6}+\frac{115^2}{4}+\frac{183^2}{6}-\frac{683^2}{21}=150.455$$

Residual sum of squares, by subtraction=307.238−150.455=156.783

**Table 7.2** Analysis of variance table for the Alloys Example

| Source | Sum of squares | Degrees of freedom | Mean square | $F$ ratio |
|---|---|---|---|---|
| Treatments | 150.455 | 3 | 50.15 | 5.44 |
| Residual (by subtraction) | 156.783 | 17 (=20−3) | 9.22 | |
| Total | 307.238 | 20 | | |

The $F$ ratio we have obtained allows us to answer the question of whether the treatment means are significantly different from each other or not. Assume the null hypothesis that all the population treatment means are the same. It follows that the variability between the sample means should be the same as that within the four samples themselves. The two terms on the right-hand side of (7.4) should therefore be contributing to the total sum of squares in proportion to the numbers of degrees of freedom they possess. Hence the $F$ ratio ought to be equal to 1 if the null hypothesis is correct.

The value of 5.44 we have found for the alloys data has 3 and 17 degrees of freedom and is highly significant. **We conclude that the average breaking strains of the 4 alloys are not all the same.**

Before leaving the ANOVA table to investigate the means themselves more closely, it may help the reader to understand the nature of the residual sum of squares by giving an alternative derivation of it. Making the 'error' the subject of equation (7.3) tells us that

$$\text{error} = x - \bar{x}_{\text{treatment}} \tag{7.6}$$

Squaring and summing this expression for all the $x$ values within the same treatment thus yields the corrected sum of squares for that treatment. These can be calculated separately for each of the four samples and are 21.200, 79.333, 38.750 and 17.500 respectively. Adding them together produces the residual sum of squares previously obtained by subtraction, 156.783. For this reason the treatments sum of squares is referred to as the **between-treatments sum of squares** while the residual is the **within-treatments sum of squares**. The residual sum of squares is here seen to be the total squared deviation when each sample is assessed against its own individual average. The residual number of degrees of freedom is similarly the sum of the separate within-sample degrees of freedom.

*Question 2* One of the objectives of the experiment was to investigate the difference between Irongrip and Coldsteel. Test the observed difference for significance.

*Solution* Table 7.1 includes the mean breaking strain for each of the 4 alloys. Having established that they are not simply random variations of their overall average we can make comparisons between them. A **pre-determined comparison** between two means can be achieved using a two-sample $t$ test as described in Chapter 6 with a slight modification. The pooled standard deviation $s_p$ in equation (6.9) is taken to be the square root of the residual mean square from the ANOVA table. This provides a better estimate of the background variability than equation (6.8) as it is based on all the samples rather than on just two. Of course if there is a doubt that the equal within-sample variance assumption of classical sampling theory is valid, then a straightforward two-sample $t$ test can be performed.

Under the null hypothesis that Irongrip and Coldsteel have the same population breaking strain,

$$t = \frac{34.60 - 30.50}{\sqrt{9.22}\sqrt{(\frac{1}{5} + \frac{1}{6})}} = 2.230 \tag{7.7}$$

The number of degrees of freedom of this statistic is that of the residual mean square. The 5% two-tail critical value of $t$ with 17 degrees of freedom is 2.110, and so **the mean breaking strain of Irongrip and Coldsteel are different at the 5% level of significance.**

*Question 3* Compare Irongrip and Brassmeld as a group of alloys with Tuffstuff and Coldsteel.

*Solution* Comparisons of groups of sample means with each other or of several samples with a single control sample can be made using the concept of a **contrast**. This is a linear combination of the treatment averages which has the null hypothesis expectation of zero:

$$L=\lambda_1\bar{x}_1+\lambda_2\bar{x}_2+\lambda_3\bar{x}_3+\cdots+\lambda_m\bar{x}_m \tag{7.8}$$

with

$$\lambda_1+\lambda_2+\lambda_3+\cdots+\lambda_m=0 \tag{7.9}$$

The symbol $\lambda$ is the Greek letter 'lambda'.

A contrast is a random variable as it depends on the treatment means and has standard deviation:

$$s_L=\sqrt{\left[(\text{residual mean square})\left(\frac{\lambda_1^2}{n_1}+\frac{\lambda_2^2}{n_2}+\frac{\lambda_3^2}{n_3}+\cdots+\frac{\lambda_m^2}{n_m}\right)\right]} \tag{7.10}$$

The ratio $L/s_L$ has the $t$ distribution with the same number of degrees of freedom as the residual mean square.

In order to gain familiarity with the idea of a contrast we can recalculate the $t$ statistic of the previous question by this method. The comparison between Irongrip and Coldsteel, treatments 1 and 4, is effected by putting $\lambda_1$ equal to $+1$, $\lambda_4$ equal to $-1$ and the other two values of $\lambda$ equal to zero:

$$L=1.\bar{x}_1+0.\bar{x}_2+0.\bar{x}_3+(-1)\bar{x}_4=34.60-30.50=4.10 \tag{7.11}$$

Notice that the chosen $\lambda$ values add up to zero in accordance with (7.9) and we have obtained the numerator of the $t$ statistic (7.7). The standard deviation, equation (7.10), is

$$s_L=\sqrt{\left[9.22\left(\frac{1^2}{5}+\frac{0^2}{6}+\frac{0^2}{4}+\frac{(-1)^2}{6}\right)\right]}=1.84 \tag{7.12}$$

This is the denominator of (7.7) and so the resulting $t$ statistic $(L/s_L)$ is the same as the one found before, 2.23.

Returning to the comparison of groups of treatment means, we can define the contrast

$$L=1.\bar{x}_1+1.\bar{x}_2+(-1)\bar{x}_3+(-1)\bar{x}_4=34.60+35.33-28.75-30.50$$
$$=10.68 \tag{7.13}$$

Its standard deviation is

$$s_L=\sqrt{\left[9.22\left(\frac{1^2}{5}+\frac{1^2}{6}+\frac{(-1)^2}{4}+\frac{(-1)^2}{6}\right)\right]}=2.69 \tag{7.14}$$

Hence the $t$ statistic $(L/s_L)$ is (10.68/2.69), which is 3.97. This is highly significant in a two-tail test with 17 degrees of freedom and so **the group 'Irongrip and Brassmeld' form a significantly different set of alloys from the group 'Tuffstuff and Coldsteel'.**

Another method of assessing a contrast for significance is to incorporate its effect in the analysis of variance table. The deviation of the contrast from zero can be thought of as a contribution to the treatment sum of squares and tested as such. The amount of the treatment sum of squares explained by a contrast is

$$\frac{(\lambda_1\bar{x}_1+\lambda_2\bar{x}_2+\lambda_3\bar{x}_3+\cdots+\lambda_m\bar{x}_m)^2}{\left(\frac{\lambda_1^2}{n_1}+\frac{\lambda_2^2}{n_2}+\frac{\lambda_3^2}{n_3}+\cdots+\frac{\lambda_m^2}{n_m}\right)} \tag{7.15}$$

This has one degree of freedom and when divided by the residual mean square to form an $F$ ratio gives the square of the $t$ statistic we have been using above. Once again we see that an $F$ ratio with 1 degree of freedom in the numerator is equivalent to a $t$ statistic. Contrasts are used later in the Bread Example. We finish here with a question frequently asked by researchers after an analysis of variance has been performed.

*Question 4* As Brassmeld gave the strongest sample and Tuffstuff the weakest, are they significantly different from each other?

*Solution* It is tempting to focus attention on those treatment means which have the largest difference between them. There is a subtle danger in being led by the sample values into tests of significance which were not planned before the analysis. The maximum difference between all possible pairs of means has a higher probability of being large than the difference between two means designated for testing before the data was collected. Selecting which samples to compare in a *post hoc* fashion thus distorts the level of significance at which conclusions can be drawn. Questions 2 and 3 above dealt with predetermined comparisons and the problem of self-selecting differences did not arise.

Methods for assessing the ratio of any contrast chosen in any way to its standard deviation have been suggested by Tukey for samples of equal size and by Scheffé for those of unequal size. Tukey's technique is described in (21), although in the Postmen Example of this chapter we shall use a different approach. For the alloys, we compare the ratio of the contrast

$$L=0\bar{x}_1+1\bar{x}_2+(-1)\bar{x}_3+0\bar{x}_4=35.33-28.75=6.58 \tag{7.16}$$

and its standard deviation

$$s_L=\sqrt{\left[9.22\left(\frac{0^2}{5}+\frac{1^2}{6}+\frac{(-1)^2}{4}+\frac{0^2}{6}\right)\right]}=1.96 \tag{7.17}$$

with the critical value $\sqrt{[(m-1)F]}$ derived by Scheffé. Here, $m$ is the number of treatments and $F$ has $(m-1)$ and $\nu_{\text{residual}}$ degrees of freedom, where $\nu_{\text{residual}}$ is the number of degrees of freedom of the residual sum of squares in the ANOVA table. We calculate this critical value to be, for a two-tail test at the 5% level, $\sqrt{[(4-1)(3.20)]}$, which is 3.10. This is not equal to the critical $t$ value of the solution to Question 3 because it is designed to test any contrast the researcher proposes after the analysis. It is therefore bigger and hence more demanding. The ratio of $L$ to $s_L$ is larger than 3.10 and so the difference is significant. **Brassmeld and Tuffstuff are essentially different alloys.** Other methods of "following up" an analysis of variance are illustrated in this chapter and the next one.

## The Kruskal–Wallis test

Before continuing with the analysis of variance of metric variables when there is more than one factor present, we consider an example containing ordinal data. The Kruskal–

Wallis test is a nonparametric form of a one-way analysis of variance and applies to several samples of data whose relative rather than absolute values are meaningful. In the following example a metric level of measurement cannot be reached and ranks are used.

**The road accidents example**

The severity of a road accident cannot be easily quantified. Police officers who had been at the scene of 20 accidents were asked to rank them in order of severity with 1 corresponding to the most serious. The results are:

| Location of accident | Ranks from 1 to 20 |
|---|---|
| Motorway | 1, 2, 5, 6, 7, 11, 15, 16, 19 |
| Major road | 3, 8, 10, 12, 14, 20 |
| Minor road | 4, 9, 13, 17, 18 |

We analyse the data to decide whether the 3 types of roads differ significantly in the severity of the accidents which occur on them.

*Solution* The sums of the ranks for each treatment are 82 for 'Motorway', 67 for 'Major road' and 61 for 'Minor road'. As mentioned in the Teaching Example of Chapter 4 the sum of $N$ ranks is $N(N+1)/2$ and we can check our arithmetic. In this case, $82+67+61$ gives the same result as $20(21)/2$ and so we proceed.

The null hypothesis for the test is that the treatments are the same and so each sample should have a total proportional to its size as the ranks are distributed uniformly amongst them. The deviations from these expected values can be measured by the $\chi^2$ statistic provided there are more than about 15 data values altogether:

$$\chi^2 = \frac{12}{N(N+1)}\left(\text{sum of}\left(\frac{R_i^2}{n_i}\right)\right) - 3(N+1) \tag{7.18}$$

where $R_i$ is the sum of the ranks for sample $i$ and $n_i$ is its size. The quantity $N$ is the total number of data values and the number of degrees of freedom for the statistic is the number of treatments minus 1. For the road accident data:

$$\chi^2 = \frac{12}{20(21)}\left(\frac{82^2}{9}+\frac{67^2}{6}+\frac{61^2}{5}\right) - 3(21) = 0.985 \tag{7.19}$$

This is not significantly large at the 5% level as the critical $\chi^2$ value for 2 degrees of freedom is 5.9915. **We conclude that there is no difference between the 3 types of road in the severity of the accidents occurring.** If the test just performed had indicated a pronounced lack of uniformity between the treatments then a two-sample nonparametric technique, like the Mann–Whitney test of Chapter 6, could be employed to compare them in pairs.

## Two-way analysis of variance without replication

The previous two examples each had one factor of interest which generated several samples of data because of its various levels. With two or more factors the situation becomes complicated by possible **interactions** between the factors. In the next chapter these and other features of experimental design are considered. We restrict ourselves here to the simplest cases of two-factor data structures as they often occur in practice and also serve as preparation for later work. In the first of these examples there is just one data value for each treatment and the extent of the interaction effect cannot be judged. In the second example the two-factor treatment combinations are replicated and an estimate of interaction can be made.

**The postmen example**

A sorting office has 5 postmen and there are 3 'walks' to be done every day in order to deliver the mail. Each postman is timed on each walk:

| | Time taken in minutes | | | | |
|---|---|---|---|---|---|
| Postman | Phil | Pat | Tom | Stan | Morris |
| Walk 1 | 136 | 132 | 134 | 129 | 138 |
| Walk 2 | 131 | 128 | 131 | 133 | 130 |
| Walk 3 | 147 | 141 | 140 | 130 | 136 |

*Question 1* Carry out an analysis of variance to determine the role played by the two factors 'postman' and 'walk'.

*Solution* There is only one measurement for each combination of the two factors. This lack of information means that we cannot estimate any interaction effect and use the model:

$$x = \mu + \alpha_{\text{postman}} + \beta_{\text{walk}} + \text{error} \tag{7.20}$$

**Table 7.3** Analysis of variance calculations for the Postmen Example

| | Postman | | | | | | |
|---|---|---|---|---|---|---|---|
| | Phil | Pat | Tom | Stan | Morris | Total | Mean, $\bar{x}_{\text{walk}}$ |
| Walk 1 | 16 | 12 | 14 | 9 | 18 | 69 | 13.8 |
| Walk 2 | 11 | 8 | 11 | 13 | 10 | 53 | 10.6 |
| Walk 3 | 27 | 21 | 20 | 10 | 16 | 94 | 18.8 |
| Total | 54 | 41 | 45 | 32 | 44 | 216 | |
| Mean, $\bar{x}_{\text{postman}}$ | 18.00 | 13.67 | 15.00 | 10.67 | 14.67 | | $\bar{x}=\frac{216}{15}=14.4$ |

$$\text{Sum of } (x^2)=3502;\ CS_{xx}=3502-\frac{216^2}{15}=391.60$$

$$CS_{\text{walks}}=\frac{69^2}{5}+\frac{53^2}{5}+\frac{94^2}{5}-\frac{216^2}{15}=170.80$$

$$CS_{\text{postmen}}=\frac{54^2}{3}+\frac{41^2}{3}+\frac{45^2}{3}+\frac{32^2}{3}+\frac{44^2}{3}-\frac{216^2}{15}=83.60$$

The equation corresponding to (7.3) is

$$x - \bar{x} = (\bar{x}_{\text{postman}} - \bar{x}) + (\bar{x}_{\text{walk}} - \bar{x}) + \text{error} \tag{7.21}$$

The terms in brackets are estimates of $\alpha_{\text{postman}}$ and $\beta_{\text{walk}}$. The corrected sums of squares for the postman and walk effects are calculated in Table 7.3. The figures have been reduced by 120 to make them more manageable. Addition, subtraction, multiplication or division of all the data values by the same number does not affect the resulting mean squares in an analysis of variance.

The method of the calculation is a straightforward extension of that for Table 7.1 with row totals, corresponding to 'walks', contributing to the treatments sum of squares as well as column ones. Equation (7.5) is used on rows and columns separately. As in the Alloys Example the treatment sums of squares are transferred to an ANOVA table and the residual determined by subtraction. This is shown in Table 7.4. The $F$ ratios for the two-factor effects are calculated from the residual mean sum of squares. As usual it measures the similarity between the variability amongst treatments and that of the population as a whole. The bigger it is the stronger the treatment effect.

**Table 7.4** Analysis of variance table for the Postmen Example

| Source | Sum of squares | Degrees of freedom | Mean square | *F* ratio |
|---|---|---|---|---|
| Walks | 170.80 | 2 | 85.40 | 4.98 |
| Postmen | 83.60 | 4 | 20.90 | 1.22 |
| Residual (by subtraction) | 137.20 | 8 | 17.15 | |
| Total | 391.60 | 14 | | |

Testing the $F$ ratio for 'walks' with the 5% value for 2 and 8 degrees of freedom, 4.46, we find that **there is a significant difference between the walk times and 'walk' is a significant factor**. The $F$ ratio for 'postmen', however, does not reach the 5% value of 3.84 corresponding to 4 and 8 degrees of freedom. **Hence the postmen are not fundamentally different from each other in their times.**

Having decided that the postmen factor is nonsignificant, we can re-analyse the data treating 'walk' as the only factor. Essentially this is simply merging the sum of squares due to 'postmen' with the residual and is illustrated in Table 7.5. It is similar to Table 7.2 for the Alloys Example and because of the change in the residual the $F$ ratio for 'walks' has altered slightly. It is still significant. In practice reducing the analysis by amalgamating insignificant factors with the residual does not affect the conclusions very often. As the sums of squares being combined have been shown to be similar by the $F$ test the outcome is almost the same as it was before. The reduction of the table is desirable because the residual mean square is now based on more degrees of freedom and is therefore more accurate.

**Table 7.5** Reduced ANOVA table for the Postmen Example

| Source | Sum of squares | Degrees of freedom | Mean square | *F* ratio |
|---|---|---|---|---|
| Walks | 170.80 | 2 | 85.40 | 4.64 |
| Residual (by subtraction) | 220.80 | 12 | 18.40 | |
| Total | 391.60 | 14 | | |

*Question 2* Which of the three walks are significantly different from each other?

*Solution* In the Alloys Example we used the $t$ distribution to make predetermined comparisons between treatment means and Scheffé's method for examining the maximum difference or any other comparison suggested by the analysis. Each of the samples for the 3 walks in the present example has the same size and a technique due to Tukey is preferable to Scheffé's in this case. Tukey's method is less conservative than Scheffé's and acknowledges the 'balancing' of the design that equal sample sizes provides. However, a more popular way of comparing all possible differences involves the $t$ statistic.

For samples with equal sizes the denominator in equation (7.7), which is derived from (6.9), will be the same no matter which pair of means is tested. It can be multiplied by the appropriate 5% critical $t$ value to give the **least significant difference** between any two sample means. Bearing in mind the danger of 'data dredging' discussed in the Alloys Example, such an approach should be used very carefully.

There are 5 data values in each 'walk' sample and the best estimate we have of their standard deviation is the square root of the residual mean square in Table 7.5. Making the numerator of the $t$ statistic (7.7) the subject with these values gives:

$$\text{least significant difference} = 2.179\sqrt{[18.40(\tfrac{1}{5}+\tfrac{1}{5})]} = 5.91 \qquad (7.22)$$

In general this will be

$$\text{least significant difference} = t\sqrt{\left[(\text{residual mean square})\left(\frac{2}{n}\right)\right]} \qquad (7.23)$$

The number of degrees of freedom of the $t$ statistic is the same as that of the residual mean square and $n$ is the common sample size. If the postmen factor had been significant, then the residual from Table 7.4 would be used at this stage.

Expression (7.23) represents the smallest difference between the means which signifies that two treatments are not the same. Arranging the means of the 3 'walks' from Table 7.3 in numerical order and applying this criterion we find that **walks 2 and 3 are significantly different and the mean time for walk 1 does not differ markedly from either of them**. An equivalent method of comparing the means is to calculate, say, 95% confidence intervals for each treatment population mean. In cases where these do not overlap the means are significantly different. The reader may care to verify that an analysis based on Scheffé's method gives the same result. Clearly the discussion of the differences between the means is unaffected by the subtraction of 120 from all the data values initially. The true mean times for the 3 walks, however, are 120 plus those calculated in Table 7.3.

## Two-way analysis of variance with replication

The final example of this chapter contains a two-factor data structure in which each treatment combination has been replicated. The extra amount of information this conveys about the factors has the consequence that we can estimate the degree of interaction between them. The number of replications for each treatment is the same. If it is not, then the analysis becomes more complicated and is beyond the scope of this book. A three-way analysis of variance is included in the next chapter.

**The bread example**

A baker is investigating the relationship between the quality of the bread he produces and the flour and baking time used. He makes 3 loaves of bread for each treatment combination of flour and baking time. The loaves are graded according to a quality/marketability index he has devised for the purpose. We shall treat this measurement as a metric variable and the values are:

| | Baking time (minutes) | | | |
|---|---|---|---|---|
| Flour | 25 | 30 | 35 | 40 |
| Granary | 3.7<br>4.2<br>2.9 | 4.2<br>3.8<br>4.1 | 4.1<br>4.3<br>4.2 | 4.8<br>5.5<br>5.4 |
| Wholemeal | 4.1<br>4.5<br>4.6 | 4.4<br>4.6<br>3.9 | 4.9<br>4.4<br>4.5 | 5.4<br>5.2<br>4.9 |
| Stoneground | 4.1<br>4.2<br>3.8 | 4.4<br>4.8<br>4.0 | 4.7<br>5.0<br>5.2 | 4.8<br>4.6<br>5.1 |

*Question 1* Perform an analysis of variance to determine the role of the factors.

*Solution* The analysis of variance model for a two-factor data structure with interaction is

$$x = \mu + \alpha_{\text{flour}} + \beta_{\text{time}} + \gamma_{\text{flour}\times\text{time}} + \text{error} \tag{7.24}$$

This can be written as

$$x - \bar{x} = (\bar{x}_{\text{flour}} - \bar{x}) + (\bar{x}_{\text{time}} - \bar{x}) + (\bar{x}_{\text{flour}\times\text{time}} - \bar{x}_{\text{flour}} - \bar{x}_{\text{time}} + \bar{x}) + \text{error} \tag{7.25}$$

where the terms in brackets are estimates of $\alpha_{\text{flour}}$, $\beta_{\text{time}}$ and $\gamma_{\text{flour}\times\text{time}}$ respectively. A multiplication sign denotes the interaction between factors. For instance $\bar{x}_{\text{wholemeal}\times 35\text{ minutes}}$ is the average of the data in the **cell** corresponding to bread made of wholemeal flour and baked for 35 minutes. The calculation of the relevant sums of squares to examine this model is shown in Table 7.6. In each corrected sum the treatment totals are squared and divided by the number of data values they contain. This is in accordance with equation (7.5).

Table 7.7 is the analysis of variance table. The interaction sum of squares is found by subtracting the 'flours' and 'times' ones from the overall treatments sum. The residual is likewise found by subtracting the overall treatments sum from the total, $CS_{xx}$. The number of degrees of freedom for each treatment is the number of treatments minus 1 and those of the interaction and residual are calculated by subtraction.

The table shows that the total variability, 10.716, of the 36 data values about their mean can be explained by the factors 'flour' and 'time', which combine to produce 12 separate treatments, and a residual. The treatment sum of squares, 8.136, can in turn be expressed as that due to 'flour', 0.844, that due to 'time', 5.801, and that due to an interaction between the factors, 1.491. Algebraically this breakdown of the total sum of squares agrees with the terms in the model (7.25) and measures the significance of

**Table 7.6** Analysis of variance calculations for the Bread Example

| Flour | Baking time (minutes) 25 | 30 | 35 | 40 | Total | Mean $\bar{x}_{flour}$ |
|---|---|---|---|---|---|---|
| Granary | 3.7, 4.2, 2.9 } 10.8 | 4.2, 3.8, 4.1 } 12.1 | 4.1, 4.3, 4.2 } 12.6 | 4.8, 5.5, 5.4 } 15.7 | 51.2 | 4.27 |
| Wholemeal | 4.1, 4.5, 4.6 } 13.2 | 4.4, 4.6, 3.9 } 12.9 | 4.9, 4.4, 4.5 } 13.8 | 5.4, 5.2, 4.9 } 15.5 | 55.4 | 4.62 |
| Stoneground | 4.1, 4.2, 3.8 } 12.1 | 4.4, 4.8, 4.0 } 13.2 | 4.7, 5.0, 5.2 } 14.9 | 4.8, 4.6, 5.1 } 14.5 | 54.7 | 4.56 |
| Total | 36.1 | 38.2 | 41.3 | 45.7 | 161.3 | |
| Mean, $\bar{x}_{time}$ | 4.01 | 4.24 | 4.59 | 5.08 | | $\bar{x}=4.48$ |

$$\text{Sum of } (x^2)=733.43; \qquad CS_{xx}=733.43-\frac{(161.3)^2}{36}=10.716$$

$$CS_{treatments}=\frac{(10.8)^2}{3}+\frac{(12.1)^2}{3}+\frac{(12.6)^2}{3}+\cdots-\frac{(161.3)^2}{36}=8.136$$

$$CS_{flours}=\frac{(51.2)^2}{12}+\frac{(55.4)^2}{12}+\frac{(54.7)^2}{12}-\frac{(161.3)^2}{36}=0.844$$

$$CS_{times}=\frac{(36.1)^2}{9}+\frac{(38.2)^2}{9}+\frac{(41.3)^2}{9}+\frac{(45.7)^2}{9}-\frac{(161.3)^2}{36}=5.801$$

each one of them. In the last example there was no replication within treatments and the treatments sum of squares was also the total sum of squares. We had to assume that there was no interaction and its sum of squares was due to residual error so that we could test the factor effects against it.

The $F$ ratios of the mean squares are all calculated with respect to the residual mean square. They measure the similarity between the various treatment variabilities and

**Table 7.7** Analysis of variance for the Bread Example

| Source | Sum of squares | Degrees of freedom | Mean square | $F$ ratio |
|---|---|---|---|---|
| Flours | 0.844 | 2 | 0.422 | 3.91 |
| Times | 5.801 | 3 | 1.934 | 17.91 |
| Interaction (by subtraction) | 1.491 | 6 | 0.249 | 2.31 |
| Treatments | 8.136 | 11 | 0.740 | 6.85 |
| Residual (by subtraction) | 2.580 | 24 | 0.108 | |
| Total | 10.716 | 35 | | |

the background error variance. Each ratio is tested against the appropriate critical $F$ value, for instance the interaction effect is not significant at the 5% level because 2.31 is smaller than 2.51, which is the 5% $F$ value with 6 and 24 degrees of freedom. The interaction should be tested first. If it is significant then every treatment combination represents a separate population and the analysis can be continued as for the Straight Line example of the next chapter. If the interaction is not significant then the factor effects are independent and superimposed additively onto each other. Its sum of squares can then be merged with the residual but in practice it makes little difference to the conclusions, which in this case are that **the quality of the bread is affected by both the flour used and the time it is baked**.

*Question 2* How does the average quality level of the bread vary with the baking time?

*Solution* We have seen in earlier examples how the analysis of variance can be followed up by examining the differences between the treatment means. Although those methods could be applied here we explore the property of the data that one of the factors is numerically valued. Furthermore it is a highly significant factor. A contrast can be constructed which tests the validity of a regression of the bread quality on the baking time by means of **orthogonal polynomials**.

Imagine a number scale on which the baking times are symmetrically placed with zero half-way between the middle two of them. The four values can be thought of as corresponding to −3, −1, +1, and +3 on this scale and are equally spaced at a distance of 2 apart. If there were an odd number of treatments we would call the central one zero and assign the values ±1, ±2, ±3, ... to the others. Fitting a straight line through the data values replaced by their treatment means is equivalent to considering a contrast with these coded times as $\lambda$ values:

$$L=(-3)\bar{x}_1+(-1)\bar{x}_2+(+1)\bar{x}_3+(+3)\bar{x}_4 \tag{7.26}$$

Equation (7.15) gives the contribution of a contrast to the treatment sum of squares. Applying it here we find that:

$$\text{Linear contribution}=\frac{((-3)(36.1/9)+(-1)(38.2/9)+(+1)(41.3/9)+(+3)(45.7/9))^2}{\left(\dfrac{(-3)^2}{9}+\dfrac{(-1)^2}{9}+\dfrac{(+1)^2}{9}+\dfrac{(+3)^2}{9}\right)}$$

$$=5.653 \tag{7.27}$$

In fact this is a regression sum of squares as defined in Chapter 4 and has one degree of freedom. The means in (7.27) have been recalculated to a greater accuracy than shown in Table 7.6. This is advisable as the contribution of a contrast is often small and rounding errors can seriously affect its value.

The amount of variability, 5.653, explained by the linear regression is a component of the baking times sum of squares, 5.801. We see that it accounts for almost all the effect of the baking times on the quality of the bread. It seems that the longer the loaves are in the oven the better the finished product. We shall test this shortly.

In some situations a linear model of the effect of a factor may not be adequate. A second order or quadratic dependence can be assumed by squaring the coded baking times and subtracting their average from them before using them as $\lambda$ values. This correction is necessary as all the $\lambda$ values in a contrast must add up to zero.

We found that the coded times were −3, −1, +1 and +3. Squaring them gives 9, 1, 1, and 9 with mean value (20/4), which is 5. Subtracting 5 from each square we obtain

+4, −4, −4 and +4. As $\lambda$ values can be multiplied or divided by the same number without altering the contribution of the contrast, we can take these to be +1, −1, −1, and +1. These constitute an **orthogonal contrast** to the previous linear one as the products of corresponding $\lambda$ values have zero sum. This will always be the case if linear and quadratic contrasts are constructed in this manner and all the samples have the same size. This orthogonality allows us to examine the linear and quadratic effects together.

Using (7.15) again the quadratic component is:

$$\text{Quadratic contribution} = \frac{((+1)(36.1/9)+(-1)(38.2/9)+(-1)(41.3/9)+(+1)(45.7/9))^2}{\left(\frac{(+1)^2}{9}+\frac{(-1)^2}{9}+\frac{(-1)^2}{9}+\frac{(+1)^2}{9}\right)}$$

$$= 0.147 \qquad (7.28)$$

Table 7.8 is the augmented analysis of variance table incorporating these linear and quadratic effects. As usual the mean squares are compared with the residual and the $F$ ratio for the linear effect is highly significant. The quadratic effect is not significantly large. **We conclude that there is a strong linear dependence of bread quality on the time it is baked.**

**Table 7.8** ANOVA table for the Bread Example including linear and quadratic effects

| Source | Sum of squares | Degrees of freedom | Mean square | *F* ratio |
|---|---|---|---|---|
| Flours | 0.844 | 2 | 0.422 | 3.91 |
| Times: | | | | |
| Linear | 5.653 | 1 | 5.653 | 52.34 |
| Quadratic | 0.147 | 1 | 0.147 | 1.36 |
| Higher order (by subtraction) | 0.001 | 1 | 0.001 | 0.01 |
| Times total | 5.801 | 3 | 1.934 | 17.91 |
| Interaction (by subtraction) | 1.491 | 6 | 0.249 | 2.31 |
| Treatments | 8.136 | 11 | 0.740 | 6.85 |
| Residual (by subtraction) | 2.580 | 24 | 0.108 | |
| Total | 10.716 | 35 | | |

Whilst the methods presented above can be used to find $\lambda$ values for linear and quadratic contrasts, those for higher-order polynomials are more complicated to calculate. They are quoted in tables of orthogonal polynomials such as those contained in reference (9) of the Bibliography.

## Summary

**1** Several samples of the same attribute variable can be dealt with as a contingency table. A Kruskal–Wallis analysis of ranks enables us to assess samples of ordinal data for homogeneity.

**2** The most common method for examining many samples of a single metric variable is the analysis of variance, abbreviated to ANOVA. A one-way data structure relates to a single factor with treatment replications which can be unequal in number. The total variability is broken down, by applying a model, into 'between treatments' and 'within treatments' components. The relative sizes of the two variances obtained are compared by an $F$ test. A 'follow up' appraisal of the treatment means can be achieved by predetermined two-sample $t$ tests, Scheffé's method for contrasts or a least significant difference approach. Care is necessary in testing differences because they happen to be large for the specific data in hand. The maximum difference between all possible pairs of means does not have the same probability distribution as a predetermined difference.

**3** If there are no replications in a two-factor data structure then any interaction between the factors must be ignored and assumed to be absent. There is insufficient data in that case to estimate an interaction term in the model. However, when each treatment is replicated there are cells of data values for each factor combination. The treatments sum of squares is no longer equal to the total sum of squares and the interaction component is distinct from the residual. If the interaction is significant, then the model reduces to expressing each $x$ value as a cell mean plus random error. If it is not, then the relevant term in the model is taken to be zero and the various factor effects are additive. Contrasts provide a way of investigating the behaviour of groups of treatment means and linear, quadratic and higher-order polynomial dependencies of the data on factors with numerically valued levels.

## Further reading

The analysis of variance is a fundamental tool of practical statistics. The next chapter develops it into a powerful method of examining data from a wide variety of experimental situations. It is also extended to deal with more than two factors and their interactions. All statistics texts except the most elementary contain relevant material, (9), (12), (21), (44) and (45) being particularly good. However, the non-mathematical reader should beware the heavy algebra and tortuous notation used.

Books under the heading 'Nonparametric Methods' should be consulted for exact probability techniques for the Kruskal–Wallis test with small samples. There are other nonparametric analyses of variance, and (33) has worked examples and is fairly easy to follow.

The adoption of the additive ANOVA model implies various assumptions about the factors and treatments being studied. These are discussed in (44), which also covers the problem of data with missing values. Such deficiencies can sometimes be rectified by using the model to estimate measurements which were never made. Needless to say it is a complicated procedure and raises all sorts of problems about the validity of subsequent conclusions.

## Exercises

**1** Carry out an analysis of variance on the rats' times given in Question 2 of the Exercises to Chapter 6. Verify that the $F$ ratio is the two-sample $t$ statistic squared, showing that for two samples the methods are equivalent.

**2** Three groups of cross-country cyclists were subjected to different training programmes. After some weeks they were timed on the same route and the results in minutes were as follows:

| Training programme | Times in minutes |
|---|---|
| Eatwell | 85.0, 88.2, 86.6, 86.0, 93.7, 91.3 |
| Jerkaround | 88.6, 93.1, 94.2, 89.7, 90.6, 85.3, 91.2 |
| Sleepalot | 94.2, 96.1, 92.8, 93.5, 91.4 |

(i) Subtract 80 from each of the data values and perform an analysis of variance to determine whether there are significant differences between the average times. (ii) Compare the training programmes 'Eatwell' and 'Jerkaround' with each other. (iii) Construct and test a suitable contrast to compare 'Eatwell' with the average of 'Jerkaround' and 'Sleepalot'. (iv) Use Scheffé's method to comment on the maximum difference between the treatment means.

**3** A manufacturer of electronic components made 4 types of resistors at 5 different temperature settings of his machine. The electrical resistance of each one was subsequently measured and the deviations from the desired value, in hundreds of ohms, were found to be as follows:

| Deviations in hundreds of ohms | Machine temperature (°C) | | | | |
|---|---|---|---|---|---|
| Type of resistor | 10 | 15 | 20 | 25 | 30 |
| Voltdown | 24 | 7 | 21 | 11 | 1 |
| Ampa | 21 | 12 | 9 | 10 | −8 |
| Pluggit | 4 | 12 | 7 | −3 | −14 |
| Loadup | 10 | 18 | 8 | −5 | −6 |

(i) Perform an analysis of variance to determine whether the types of resistor are behaving significantly differently from each other and whether the manufacturing temperatures are affecting the deviations. (ii) If appropriate, test the hypothesis that 'Voltdown' and 'Ampa' are significantly different as a group from the other two types. (iii) If the operating temperature is affecting the deviations significantly, test whether the dependence is linear, quadratic or of higher order.

**4** The lengths jumped by 4 frogs in each of 4 States of America in each of 4 months were as follows. Perform an analysis of variance to determine whether there is an interaction between the effects of the States and the months. Which factor is more important in influencing the length of the jumps?

| Jumps in metres | January | | February | | March | | April | |
|---|---|---|---|---|---|---|---|---|
| Texas | 1.00 | 0.81 | 1.32 | 0.83 | 0.90 | 0.92 | 0.91 | 1.15 |
| | 1.43 | 1.10 | 1.08 | 0.90 | 1.09 | 1.10 | 0.93 | 0.90 |
| Arizona | 1.25 | 1.25 | 1.15 | 0.68 | 1.08 | 0.88 | 1.33 | 1.15 |
| | 1.27 | 1.02 | 1.14 | 0.73 | 1.00 | 1.12 | 1.21 | 0.97 |
| Oklahoma | 1.11 | 1.02 | 1.27 | 0.98 | 0.65 | 0.75 | 1.06 | 0.98 |
| | 0.85 | 0.80 | 0.75 | 1.09 | 0.96 | 0.82 | 0.70 | 0.98 |
| Nevada | 1.38 | 1.23 | 1.15 | 1.22 | 1.05 | 1.10 | 1.13 | 1.03 |
| | 1.12 | 1.22 | 0.93 | 1.06 | 0.95 | 1.35 | 1.20 | 0.98 |

# 8 Experimental design

The links between the design and analysis stages of a statistical investigation were discussed in Chapter 1. When planning the collection of data we must be fully aware of the methods available for its subsequent analysis. Previous chapters have covered these methods and so our attention can now be directed to the principles of experimental design themselves. They have particular importance when there are several factors forming many different treatments, each with its own sample of data values. The analysis of variance techniques of the last chapter are extended and play a prominent role here as they allow us to fit statistical models to data structures designed for specific experimental situations.

The origins of experimental design methodology are in agricultural research. The testing of new crops, fertilisers and insecticides on pieces of land known to be affected by other factors requires careful thought and preparation. The term **plot** derives from this application and describes any experimental unit, for example a person, as well as a section of ground. The grouping of plots into blocks is also an agricultural concept and is discussed later in its wider context.

Certain fundamental considerations should be stated before we look at particular designs. The importance of **replication**, the repetition of a treatment on as many experimental units as possible, has been stressed elsewhere in this book. A related idea is that of **randomisation**, in which treatments are assigned to plots in a random manner. This means that every plot or group of plots has an equal probability of receiving each treatment. The procedure minimises the effects of any bias the experimenter, plots or treatments may impose on the variables being measured. A further way of reducing bias in some situations is to arrange the plots in a random sequence before making observations. An operator's gradual acclimatisation to a measuring instrument can bias the earliest readings. Randomisation is especially necessary when the same apparatus is used repeatedly in several experiments. Allocating the same machine, for example, to the same treatment introduces the factor 'machine' into the data structure in an unnecessary way. Randomisation is achieved by tossing a coin, throwing dice or performing similar trials to decide which plots receive which treatments. In clinical trials randomisation is taken one step further in **double blind** trials. Here neither the clinician nor the patient knows whether an active treatment or a **placebo**, a dummy drug, is being used.

A design is **balanced** if there is a degree of symmetry to the numbers of plots corresponding to each treatment. When there is just one factor this implies that there is the same number of replications for each level of the factor. For two factors it means having equal numbers of replicates for each combination of the factor levels. It is desirable for theoretical reasons to have some form of balancing as the tests we use are then less sensitive to non-normal parent populations.

## Blocking

Factors which are suspected of affecting the response of the experimental units to the treatments cannot always be measured or controlled. It may be possible to group the plots into **blocks** of similar items and apply each of the treatments to a different member of the same block. The blocks then constitute experiments in miniature in which all the levels of the factors of interest are tested. This method of attempting to eliminate the effects of spurious factors from an experiment appeared in Chapter 4. The matched-pair data of the various examples there illustrate blocking for a factor with just two levels. Blocks are **complete** if they contain enough plots for every treatment to be applied to them and are **randomised** if the treatments are allocated to plots within them at random. Here is an example of a complete randomised blocks experiment.

### The mice example

A new diet is to be tested on mice. There are low- and high-protein versions and the weight gains of the mice are to be compared with those on a control diet of bread and milk. In an attempt to eliminate genetic, initial weight and age factors from the experiment, a group of 3 mice was chosen from each of 8 litters. The 3 mice in each group were then fed on different diets allocated to them at random. The weight gains over the same period of time were:

| Weight gain (g) | Diet | | |
|---|---|---|---|
| Block | Control | Low protein | High protein |
| 1 | 41.7 | 42.3 | 42.8 |
| 2 | 42.7 | 43.1 | 44.5 |
| 3 | 43.5 | 44.5 | 44.0 |
| 4 | 43.1 | 43.2 | 43.5 |
| 5 | 42.8 | 43.6 | 44.6 |
| 6 | 44.5 | 44.1 | 44.3 |
| 7 | 42.9 | 43.9 | 43.2 |
| 8 | 43.0 | 44.3 | 43.2 |

*Question 1* Do the weight gains vary significantly between the 3 diets?

*Solution* The two factors here are diets and blocks. The diets are a fixed effect as their occurrence is within our control whereas the blocks are a random effect because choosing a litter could have many outcomes other than the 8 we observed. We need to consider a **mixed model**

$$x = \mu + \alpha_{\text{diet}} + \beta_{\text{block}} + \text{error} \tag{8.1}$$

with one fixed effect and one random effect. In practice blocking is often considered to be a fixed effect and as we are not primarily interested in the blocked factors it will not affect the following analysis. Random effects are generally troublesome to test and it is debatable what the results of such a test actually mean.

The calculation of the treatment and blocks sums of squares for this two-way analysis of variance is similar to that of the Postmen Example of the last chapter. The details are in Table 8.1 together with the ANOVA table.

Every data value has been reduced by 40 in order to keep the scale of the summations manageable. Such linear adjustments to the data do not alter an analysis of variance.

The $F$ ratios from Table 8.1 can be compared with the 5% significant values from the tables on page 191. **We find that both diets and blocks are significant factors.** It is

**Table 8.1** Analysis of variance for the Mice Example

| Block | Diet | | | Total | Mean |
|---|---|---|---|---|---|
| | Control | Low protein | High protein | | |
| 1 | 1.7 | 2.3 | 2.8 | 6.8 | 2.27 |
| 2 | 2.7 | 3.1 | 4.5 | 10.3 | 3.43 |
| 3 | 3.5 | 4.5 | 4.0 | 12.0 | 4.0 |
| 4 | 3.1 | 3.2 | 3.5 | 9.8 | 3.27 |
| 5 | 2.8 | 3.6 | 4.6 | 11.0 | 3.67 |
| 6 | 4.5 | 4.1 | 4.3 | 12.9 | 4.30 |
| 7 | 2.9 | 3.9 | 3.2 | 10.0 | 3.33 |
| 8 | 3.0 | 4.3 | 3.2 | 10.5 | 3.50 |
| Total | 24.2 | 29.0 | 30.1 | 83.3 | |
| Mean | 3.03 | 3.63 | 3.76 | | |

$$\text{Sum of } (x^2)=302.87; \qquad CS_{xx}=302.87-\frac{(83.3)^2}{24}=13.7496$$

$$CS_{\text{diets}}=\frac{(24.2)^2}{8}+\frac{(29.0)^2}{8}+\frac{(30.1)^2}{8}-\frac{(83.3)^2}{24}=2.4608$$

$$CS_{\text{blocks}}=\frac{(6.8)^2}{3}+\frac{(10.3)^2}{3}+\frac{(12.0)^2}{3}+\cdots+\frac{(10.5)^2}{3}-\frac{(83.3)^2}{24}=7.5563$$

| Source | Sum of squares | Degrees of freedom | Mean square | $F$ ratio |
|---|---|---|---|---|
| Diets | 2.4608 | 2 | 1.2304 | 4.62 |
| Blocks | 7.5563 | 7 | 1.0795 | 4.05 |
| Residual (by subtraction) | 3.7325 | 14 | 0.2666 | |
| Total | 13.7496 | 23 | | |

reassuring to discover that there are factors in the mice population causing the blocks to behave differently from each other. These factors were suspected as being present beforehand and were the reason that the experiment was blocked. If they had not proved signficant then a one-way analysis could be performed merging the blocks sum of squares with the residual as in the Postmen Example of Chapter 7.

*Question 2* Does the low-protein diet produce significantly bigger weight gains than the control diet?

*Solution* The possibilities for a 'follow up' examination of the means after an analysis of variance were discussed in the Alloys Example of the last chapter. When one of the treatments is a control there are obvious pre-determined comparisons to be made and here we apply the $t$ test for the equality of two population means. Equation (6.9) for a two-sample $t$ statistic with the standard deviation estimate taken to be the square root of the residual mean sum of squares is

$$t=\frac{\left(\frac{29.0}{8}\right)-\left(\frac{24.2}{8}\right)}{\sqrt{0.2666}\sqrt{\left(\frac{1}{8}+\frac{1}{8}\right)}}=2.32 \tag{8.2}$$

This has the residual degrees of freedom 14, and is significant in a one-tail test. **The low-protein diet does produce larger weight gains than the control diet.**

## Factorial designs

The Postmen and Bread Examples of the last chapter had two factors with data for every treatment combination given by the levels of those factors. Such data is in a **two-way cross classification** and the design is said to be **crossed** or **factorial.** If all possible combinations of the factor levels are present, then the experiment is **completely crossed** or has a **complete factorial design.** A factor is often the presence or absence of an attribute. The next example is an extension of the two-factor analysis of the Bread Example to the three-factor case.

### The straight-line example

A drug which affects the ability to coordinate hand and eye movements is being studied. Volunteers were given the drug and suspended upside-down by their ankles. They then attempted to draw a straight line parallel to a direction indicated on a piece of paper attached to the floor. The angles made by the lines with the marked direction for 24 subjects were:

| Angle (degrees) | Men | | Women | |
|---|---|---|---|---|
| Dose | Left-handed | Right-handed | Left-handed | Right-handed |
| Low | 32<br>12 | 22<br>16 | 25<br>20 | 11<br>3 |
| Medium | 15<br>25 | 27<br>39 | 22<br>34 | 40<br>36 |
| High | 31<br>26 | 27<br>18 | 44<br>42 | 41<br>32 |

*Question 1* Carry out a three-way analysis of variance to determine which factors and interactions are important.

*Solution* The factors 'sex' and 'dominant hand' have 2 levels each while the drug has 3 levels. The design is thus a **2×2×3 factorial.** The model assumed for factorial designs is an extension of (7.24). In this case it is:

$$x=\mu+(\text{sex})+(\text{hand})+(\text{dose})+(\text{sex}\times\text{hand})+(\text{sex}\times\text{dose})$$
$$+(\text{hand}\times\text{dose})+(\text{sex}\times\text{hand}\times\text{dose})+\text{error} \tag{8.3}$$

with the various terms indicated by the factors in brackets. The three factors are **main effects**; the two-factor interactions are **first-order interactions** while the three-factor interaction is **second order**. The parameters in the model can be estimated from averages derived from the data. The main-effect and first-order terms have similar expressions to those given in the Bread Example by equation (7.25). The second-order term is the appropriate cell mean minus the overall mean, the three main-effect terms and the three relevant first-order interaction terms. As usual in an analysis of variance, however, we proceed straight to the contributions to the total sum of squares made by each of these terms rather than to the terms themselves.

Table 8.2 shows the calculations for the analysis. The total, treatments, and main effects sums are fairly simple applications of (7.5). The first-order interaction sums are

**Table 8.2** Corrected sums of squares calculations for the Straight Line Example

| | Men | | Women | | |
|---|---|---|---|---|---|
| Dose | Left-handed | Right-handed | Left-handed | Right-handed | Total |
| Low | 32, 12 } 44 | 22, 16 } 38 | 25, 20 } 45 | 11, 3 } 14 | 141 |
| Medium | 15, 25 } 40 | 27, 39 } 66 | 22, 34 } 56 | 40, 36 } 76 | 238 |
| High | 31, 26 } 57 | 27, 18 } 45 | 44, 42 } 86 | 41, 32 } 73 | 261 |
| Total | 141 | 149 | 187 | 163 | 640 |

Sum of $(x^2) = 19\,794$; $\quad CS_{xx} = 19\,794 - \dfrac{(640)^2}{24} = 2727.333$

$$CS_{\text{treatments}} = \frac{(44)^2}{2} + \frac{(38)^2}{2} + \frac{(45)^2}{2} + \cdots + \frac{(73)^2}{2} - \frac{(640)^2}{24} = 2167.333$$

$$CS_{\text{sex}} = \frac{(141+149)^2}{12} + \frac{(187+163)^2}{12} - \frac{(640)^2}{24} = 150.000$$

$$CS_{\text{hand}} = \frac{(141+187)^2}{12} + \frac{(149+163)^2}{12} - \frac{(640)^2}{24} = 10.667$$

$$CS_{\text{dose}} = \frac{(141)^2}{8} + \frac{(238)^2}{8} + \frac{(261)^2}{8} - \frac{(640)^2}{24} = 1014.083$$

$$CS_{\text{sex}\times\text{hand}} = \frac{(141)^2}{6} + \frac{(149)^2}{6} + \frac{(187)^2}{6} + \frac{(163)^2}{6} - \frac{(640)^2}{24} - CS_{\text{sex}} - CS_{\text{hand}} = 42.667$$

$$CS_{\text{sex}\times\text{dose}} = \frac{(82)^2}{4} + \frac{(59)^2}{4} + \frac{(106)^2}{4} + \frac{(132)^2}{4} + \frac{(102)^2}{4} + \frac{(159)^2}{4} - \frac{(640)^2}{24} - CS_{\text{sex}} - CS_{\text{dose}} = 406.750$$

$$CS_{\text{hand}\times\text{dose}} = \frac{(89)^2}{4} + \frac{(52)^2}{4} + \frac{(96)^2}{4} + \frac{(142)^2}{4} + \frac{(143)^2}{4} + \frac{(118)^2}{4} - \frac{(640)^2}{24} - CS_{\text{hand}} - CS_{\text{dose}} = 503.083$$

found by ignoring the third factor. For instance, pooling the data for men and women gives Table 8.3. Each cell of this table represents four data values and so the treatments sum of squares is as shown for $CS_{\text{hand}\times\text{dose}}$ in Table 8.2. The sums of squares due to the corresponding main effects are subtracted from this to produce the interaction sum. We have really performed a two-way analysis of variance of Table 8.3 in order to obtain the interaction term.

**Table 8.3** 'Hand' versus 'dose' for the Straight Line Example

| Dose | Left-handed | Right-handed |
|---|---|---|
| Low | 89 | 52 |
| Medium | 96 | 142 |
| High | 143 | 118 |

The resulting ANOVA table with all the main effects and interactions is in Table 8.4. The second-order interaction and residual sums of squares are evaluated by subtraction. Although the numbers of degrees of freedom for the interactions can be determined by subtraction, note that they are simply the products of those for the corresponding main effects.

**Table 8.4** Analysis of variance table for the Straight-line Example

| Source | Sum of squares | Degrees of freedom | Mean square | *F* ratio |
|---|---|---|---|---|
| Sex | 150.000 | 1 | 150.000 | 3.21 |
| Hand | 10.667 | 1 | 10.667 | <1 |
| Dose | 1014.083 | 2 | 507.042 | 10.87 |
| Sex × hand | 42.667 | 1 | 42.667 | <1 |
| Sex × dose | 406.750 | 2 | 203.375 | 4.36 |
| Hand × dose | 503.083 | 2 | 251.542 | 5.39 |
| Sex × hand × dose (by subtraction) | 40.083 | 2 | 20.042 | <1 |
| Treatments | 2167.333 | 11 | 197.030 | 4.22 |
| Residual (by subtraction) | 560.000 | 12 | 46.667 | |
| Total | 2727.333 | 23 | | |

The *F* ratios in Table 8.4 are calculated with respect to the residual mean square as is appropriate for a fixed-effects model. As mentioned earlier, the analysis of random effects is more complicated and apart from their appearance in blocked experiments they are not covered in this book. Our conclusions for the ANOVA table are that **the dose and the first-order interactions between sex and dose and between hand and dose are significant**. The other *F* ratios in Table 8.4 are not sufficiently large when compared with critical values to be significant.

*Question 2* Follow up these conclusions with an analysis of the treatment means.

*Solution* Whereas we have the $t$ distribution and contrasts to help us to understand main effects, interactions are often more difficult to explain. All that can be said in general terms is that there is a departure from additivity in the model. Something about the situation is not being described adequately by the mere superposition of effects in an arithmetical fashion.

It is an unfortunate fact that transforming a variable, for instance taking square roots of data values, can alter the amount of interaction revealed by an analysis of variance. We must therefore bear in mind that just as in testing means we are making pronouncements about numerical values and not about the essential features of the factors themselves. In the present example we are talking just as much about the way dose has been measured and the sensitivity of the scale as about the concept of 'dose' itself.

A helpful way of viewing interaction effects is to construct a **table of means**. The one for 'hand and dose' is obtained from Table 8.3 by dividing each entry by the number of data values it contains, 4, to produce Table 8.5. This shows an interesting phenomenon in which the level of dose causes a gradual increase in the error of alignment of the straight lines drawn by left-handed people but a peak at the medium dose for right-handers. **It seems that right-handed folk have an ability to draw accurate lines while suspended upside down but this ability rapidly disappears when more drug is administered. Even more remarkable is that the ability tends to return when the dose level is highest!** These features explain the interaction between handedness and dose.

**Table 8.5** Table of means for the 'hand *vs* dose' interaction of the Straight-line Example

| Dose | Left-handed | Right-handed |
|---|---|---|
| Low | 22.25 | 13.00 |
| Medium | 24.00 | 35.50 |
| High | 35.75 | 29.50 |

Table 8.6 is the corresponding table of means for the other significant interaction, that between sex and dose. **Clearly the drug affects men and women differently with women drawing better lines than men at low doses but losing the ability at higher levels. Men seem more uniform in their performance.**

Having arrived at what we feel is a satisfactory explanation of the interactions, it is unnecessary to proceed and test main effects. It is hardly surprising that the factor 'dose' should prove significant when it is considered against the background provided by the last two tables of means. A $t$ statistic like equation (8.2) can be used to test pairs of means or contrasts and to derive a least significant difference if required.

**Table 8.6** Table of means for the 'sex *vs* dose' interaction of the Straight-line Example

| Dose | Men | Women |
|---|---|---|
| Low | 20.5 | 14.75 |
| Medium | 26.5 | 33.00 |
| High | 25.5 | 39.75 |

Scheffé's method could also be applied, and as before the best estimate of the standard deviation we have is the square root of the residual mean square. Given the multitude of possible pairs of means which can be calculated from Tables 8.5 and 8.6 the reader is reminded of the warning contained in the solution to Question 4 of the Alloys Example in Chapter 7. The more factors, and hence the more averages which can be computed, the more chance that two of them will be far apart due simply to random fluctuations.

## Latin square designs

The factorial design in the last example was **complete** as measurements for all possible treatments were made. This balanced state of affairs can be difficult to achieve when there are several factors with correspondingly many treatment combinations or each experimental determination is expensive or time consuming. A **fractional** or **incomplete** design has treatments missing from the data but retains certain aspects of balancing. A popular example is the Latin square design, where the levels of one factor are randomised over all possible combinations of the others. The price we pay for this reduction in the number of plots needed is that we cannot assess the extent of any interactions between the factors. In this respect it is similar to a complete factorial design with one observation per cell as in the Postmen Example of Chapter 7. Here is an example.

### The surgeons example

A patient's length of stay in hospital for a certain type of operation is believed to depend on the ward, the operating theatre used and the consultant surgeon. A hospital has 4 wards and 4 operating theatres but it was not possible to arrange for each of the 4 surgeons, Messrs Allout, Botcher, Cutsit and Diggin, to treat a patient in every ward and every theatre exactly once. Sixteen patients were allocated randomly to wards and theatres and the following schedule of surgeons prepared. The resulting lengths of stay for each patient is also shown:

| Length of stay (days) | Operating theatre | | | |
|---|---|---|---|---|
| Ward | 1 | 2 | 3 | 4 |
| Kilmore | *A*: 10 | *B*: 9 | *C*: 8 | *D*: 9 |
| Stitchem | *B*: 7 | *D*: 11 | *A*: 6 | *C*: 5 |
| Legsup | *C*: 10 | *A*: 9 | *D*: 10 | *B*: 5 |
| Betternow | *D*: 12 | *C*: 10 | *B*: 9 | *A*: 10 |

The letters refer to the initials of the surgeons' names and appear just once in every row and every column. Hence each surgeon operates on only four patients instead of all 16 as in a complete factorial design. The effect of each surgeon, however, is applied equally to every ward and every theatre. The rows and columns of a Latin square should be assigned randomly to the factor levels. In agricultural experiments the testing of a control crop in the same corner of a field whenever a new type of crop is to be assessed is obviously suspect. In fact Latin square designs are particularly suitable for partitioning a field for different treatments as any climatic or soil quality characteristics extending across the field in either direction are neutralised.

*Question 1* Determine the relative importance of the three factors.

*Solution* The analysis of variance model for a Latin square has no interaction terms and in this case is

$$x = \mu + \alpha_{\text{ward}} + \beta_{\text{theatre}} + \gamma_{\text{surgeon}} + \text{error} \quad (8.4)$$

With terms estimated from the data this becomes

$$x = \bar{x} + (\bar{x}_{\text{ward}} - \bar{x}) + (\bar{x}_{\text{theatre}} - \bar{x}) + (\bar{x}_{\text{surgeon}} - \bar{x}) + \text{error} \quad (8.5)$$

The corrected sums of squares for the three main effects are calculated from their totals as shown in Table 8.7. The row and column totals relate to wards and theatres whilst the surgeons' totals have to be determined separately. They are included in the table. All three $F$ ratios are significant against the 5% value of 4.76 for 3 and 6 degrees of freedom. **Hence all the factors are important in affecting the length of stay of a patient at the hospital. The order of importance seems to be surgeons followed by wards followed by theatres.**

**Table 8.7** Analysis of variance for the Surgeons Example

| Ward | Operating theatre | | | | Total |
|---|---|---|---|---|---|
| | 1 | 2 | 3 | 4 | |
| Kilmore | $A$: 10 | $B$: 9 | $C$: 8 | $D$: 9 | 36 |
| Stitchem | $B$: 7 | $D$: 11 | $A$: 6 | $C$: 5 | 29 |
| Legsup | $C$: 10 | $A$: 9 | $D$: 10 | $B$: 5 | 34 |
| Betternow | $D$: 12 | $C$: 10 | $B$: 9 | $A$: 10 | 41 |
| Total | 39 | 39 | 33 | 29 | 140 |

Surgeons' totals: $A=35$, $B=30$, $C=33$, $D=42$

$$\text{Sum of } (x^2) = 1288;\ CS_{xx} = 1288 - \frac{(140)^2}{16} = 63.0$$

$$CS_{\text{wards}} = \frac{(36)^2}{4} + \frac{(29)^2}{4} + \frac{(34)^2}{4} + \frac{(41)^2}{4} - \frac{(140)^2}{16} = 18.5$$

$$CS_{\text{theatres}} = \frac{(39)^2}{4} + \frac{(39)^2}{4} + \frac{(33)^2}{4} + \frac{(29)^2}{4} - \frac{(140)^2}{16} = 18.0$$

$$CS_{\text{surgeons}} = \frac{(35)^2}{4} + \frac{(30)^2}{4} + \frac{(33)^2}{4} + \frac{(42)^2}{4} - \frac{(140)^2}{16} = 19.5$$

| Source | Sum of squares | Degrees of freedom | Mean square | $F$ ratio |
|---|---|---|---|---|
| Wards | 18.5 | 3 | 6.167 | 5.28 |
| Theatres | 18.0 | 3 | 6.000 | 5.14 |
| Surgeons | 19.5 | 3 | 6.500 | 5.57 |
| Residual (by subtraction) | 7.0 | 6 | 1.167 | |
| Total | 63.0 | 15 | | |

*Question 2* Investigate the treatment means.

*Solution* The usual procedures for studying the means involving the $t$ distribution and contrasts are available to us but it is rather worrying that we have three significant main effects with no estimate of their almost certain interaction. A technique which can be used after any analysis of variance is to examine the **residual error** for each data value after that part explained by the model has been removed. The working here is in Table 8.8. The treatment means are evaluated and then for each cell equation 8.5 is applied to determine the error. For example, the entry for Betternow ward with operating theatre 3 is the result of the calculation

$$\begin{aligned} \text{error} &= x - \bar{x} - (\bar{x}_{\text{Betternow}} - \bar{x}) - (\bar{x}_{\text{ward 3}} - \bar{x}) - (\bar{x}_{\text{Botcher}} - \bar{x}) \\ &= 9 - 8.75 - (10.25 - 8.75) - (8.25 - 8.75) - (7.50 - 8.75) = +0.50 \end{aligned} \quad (8.6)$$

**Table 8.8** Residual errors for the Surgeons Example

| Ward | Operating theatre | | | | Treatment mean |
|---|---|---|---|---|---|
| | 1 | 2 | 3 | 4 | |
| Kilmore | *A*: 0.00 | *B*: +0.25 | *C*: 0.00 | *D*: −0.25 | 9.00 |
| Stitchem | *B*: 0.00 | *D*: +1.00 | *A*: −0.75 | *C*: −0.25 | 7.25 |
| Legsup | *C*: +1.00 | *A*: −0.50 | *D*: +0.25 | *B*: −0.75 | 8.50 |
| Betternow | *D*: −1.00 | *C*: −0.75 | *B*: +0.50 | *A*: +1.25 | 10.25 |
| Treatment mean | 9.75 | 9.75 | 8.25 | 7.25 | |

Surgeon means: $A$=8.75, $B$=7.50, $C$=8.25, $D$=10.5

Overall mean, $\bar{x}$=8.75

Notice that the row, column and surgeon total residual errors are all zero. This will always be the case and is a useful check on our arithmetic. It is also true that the residual sum of squares is the sum of the errors squared.

So Table 8.8 provides yet another way of raking the ashes after an analysis of variance has been performed. It shows the discrepancies between the data and the predictions given by the model. Hence large numerical values represent combinations of factors for which the model does not work particularly well: for instance **Mr Allout with a patient from Betternow ward in operating theatre 4 seems highly interactive. A patient who could choose might decide on Mr Botcher, Stitchem ward and operating theatre 2 because they have the lowest means possible.** This combination was not tested in the experiment and so no firm conclusions about its efficacy can be made.

## Nested or hierarchical designs

The factors in a factorial design are given equal status in the hypothesised model and hence in the analysis. Sometimes one factor acts as a **main group** effect while the levels of another, the **sub-group** factor, break that effect down into treatments. Such factors are **nested** one beneath the other and the experimental design is **nested** or **hierarchical. Split-plot** or **split-unit** designs form a special case of this structure in which each subject is tested with two or more treatments instead of with just one. The subject is then the main group factor with the various treatment responses within that subject constituting the sub-group factor. A hierarchical data structure also arises when it is not possible to control one factor independently of another and so a different set of levels for it is used with each level of the main group factor. For instance in the Bread Example of this chapter it may not have been possible or appropriate to bake loaves made of each type of flour for each of the preferred times. There would then be data on a different set of baking times for each flour and the two factors would be nested rather than crossed. The next example deals with only two factors but can be easily generalised to cope with more.

**The radiation example**

Two fields were chosen on each of three farms. The radiation levels at three points in each field were measured and the observations in scaled units tabulated:

| Radiation level (scaled units) | Farm | | | | | |
|---|---|---|---|---|---|---|
| | Appledene | | Butterdale | | Cowfields | |
| Field | 1 | 2 | 1 | 2 | 1 | 2 |
| Replications | 3<br>7<br>4 | 6<br>7<br>5 | 7<br>9<br>6 | 6<br>8<br>10 | 9<br>7<br>11 | 10<br>11<br>8 |

We use an analysis of variance to discover whether differences between farms or between fields within farms are significant effects.

*Solution* It would be a mistake to consider this data as a factorial design with 'farm' and 'field' as a cross-classification. The fields designated '1' have nothing in common with each other and do not represent a treatment level applied to each of the farms. The factor 'field' is not a main effect but an interaction with 'farm' and the model for a nested structure like this is

$$x = \mu + \alpha_{\text{farm}} + \beta_{\text{farm} \times \text{field}} + \text{error} \qquad (8.7)$$

The parameters can be estimated from the data as follows:

$$x = \bar{x} + (\bar{x}_{\text{farm}} - \bar{x}) + (\bar{x}_{\text{farm} \times \text{field}} - \bar{x}_{\text{farm}}) + \text{error} \qquad (8.8)$$

Table 8.9 contains the analysis of variance for this model. The total and treatments sums of squares are evaluated first. At this stage the overall significance of the treatments could be tested as for a one-way classification if required. The factor here is 'fields' and it has 6 levels.

The next step is to subdivide the treatments sum of squares into components due to differences between farms. The grouping of fields into farms should account for some of the variability between them. The corrected sum of squares for 'farms' is calculated and subtracted from the treatments sum of squares to give the contribution of the fields and farms interaction. This method of calculation reflects the hierarchical structure of the data. The lowest level, field, is processed first and the next level, farm, used to explain the differences in the means. Any variability left is attributed to 'fields within farms'. The structure could continue and if the farms were grouped in some way then their sum of squares would be further partitioned into 'between farms' and 'within farms'.

**Table 8.9** Analysis of variance for the Radiation Example

| | Farm | | | | | | |
|---|---|---|---|---|---|---|---|
| | Appledene | | Butterdale | | Cowfields | | |
| Field | 1 | 2 | 1 | 2 | 1 | 2 | |
| Replications | 3<br>7<br>4 | 6<br>7<br>5 | 7<br>9<br>6 | 6<br>8<br>10 | 9<br>7<br>11 | 10<br>11<br>8 | Total |
| Field total | 14 | 18 | 22 | 24 | 27 | 29 | 134 |
| Farm total | 32 | | 46 | | 56 | | |

Sum of $(x^2)=1086$; $\quad CS_{xx}=1086-\frac{(134)^2}{18}=88.444$

$$CS_{\text{treatments}}=\frac{(14)^2}{3}+\frac{(18)^2}{3}+\frac{(22)^2}{3}+\frac{(24)^2}{3}+\frac{(27)^2}{3}+\frac{(29)^2}{3}-\frac{(134)^2}{18}=52.444$$

$$CS_{\text{farms}}=\frac{(32)^2}{6}+\frac{(46)^2}{6}+\frac{(56)^2}{6}-\frac{(134)^2}{18}=48.444$$

| Source | Sum of squares | Degrees of freedom | Mean square | *F* ratio |
|---|---|---|---|---|
| Farms | 48.444 | 2 | 24.222 | 8.07 |
| Fields within farms (by subtraction) | 4.000 | 3 | 1.333 | <1 |
| Treatments | 52.444 | 5 | 10.489 | 3.50 |
| Residual (by subtraction) | 36.000 | 12 | 3.000 | |
| Total | 88.444 | 17 | | |

Having derived the various sums of squares which show the contributions made by each term in the model (8.8), we can test their significance. If the farms or the fields within the farms have been chosen at random, then the ANOVA table must be analysed taking into account these random effects. For simplicity we assume here that our conclusions are to be valid just for the three farms and six fields appearing in the data, hence reducing the situation to one with fixed effects. In this case the mean squares are compared with the residual mean square as in Table 8.9. We see that **the variability of fields within farms is insignificant but that occurring between farms is not.** The radiation levels between farms are therefore significantly different from each other but the levels within farms are uniform.

## Summary

**1** Replication and randomisation are important features to build into the design of experiments. They help eliminate bias in making observations. It is also desirable to balance the design in some way as the procedures for analysing complicated data structures are more sensitive to non-normality in unbalanced situations.

**2** Miscellaneous factors which we suspect influence our measurements but whose details are of no interest to us can be dealt with by blocking. If a full range of treatments is applied to each block of matched subjects, then the blocks are complete. Treatments should be assigned to members of a block randomly if possible. Although the blocks produce a random effect as far as the analysis of variance model is concerned we can still test the treatment effect against the residual in the usual way.

**3** A factorial design for two or more factors involves administering different combinations of factor levels to different samples of subjects. The design is also said to be crossed and is complete if every possible treatment is used. The treatments sum of squares in the analysis of variance is partitioned into main effects, first-order interactions between two factors, second-order interactions between three factors, and so on. When interactions are significant tables of means may help to show whereabouts the additivity of factors assumed by the model has been violated.

**4** A fractional or incomplete factorial design has some treatment combinations missing from the data structure. A Latin square design is an example of this in which one factor has relatively few replications and is in conjunction with other factor combinations which form a complete factorial structure within themselves. Nested or hierarchical designs have factors which have a logical ordering into main group effects, sub-group effects, sub-sub-group effects and so on. They are used when the levels of the sub-group factor are peculiar to the level of the main group factor and cannot be related to each other across the main group effect.

## Further reading

The Bibliography has a separate section of references on experimental design. They are all standard works but concentrate on analysis rather than the planning of an investigation from the initial stages. Topics covered in these books include models with random effects, incomplete block designs, data with missing values, designs with the effects of one factor mixed up or confounded with those of another and specialised methods for coping with large factorial designs. The reader is warned that they are highly mathematical. Many of the statistics texts written for specific disciplines cited in the Bibliography have sections on data structures, for instance references (12), (16) and (19).

## Exercises

**1** The matched-pair data for the Paint Example of Chapter 4 can be thought of as a paint factor with two levels replicated on 8 blocks of paired pieces of wood. Derive an analysis of variance table and verify that the square root of the $F$ ratio for the factor effect is equal to the $t$ value found in the example.

**2** Five students obtained the following percentages in their school examinations:

| Percentage | Student | | | | |
|---|---|---|---|---|---|
| Subject | Arthur | Boris | Charles | Dick | Ernie |
| English | 69 | 65 | 51 | 69 | 66 |
| History | 69 | 63 | 59 | 65 | 62 |
| Mathematics | 55 | 56 | 50 | 53 | 54 |
| Geography | 67 | 73 | 62 | 67 | 59 |

(i) Perform an analysis of variance treating the students as blocks to determine whether there is a significant difference between the average subject marks. (ii) Is there a significant difference between the average marks in English and Geography? (iii) Ignore the blocking and analyse the data as a one-way structure with the five students as unrelated replications. Are the conclusions affected?

**3** The thrusts produced by five rocket engines on five different fuels and five different injection settings were measured:

| Thrust (thousand kg) | Injection rate (litres/second) | | | | |
|---|---|---|---|---|---|
| Fuel | 1.2 | 1.5 | 1.6 | 1.9 | 2.5 |
| Blastout | *A*: 4.6 | *B*: 4.7 | *C*: 5.1 | *D*: 4.3 | *E*: 5.1 |
| Launcher | *C*: 5.3 | *D*: 4.8 | *E*: 5.1 | *A*: 4.9 | *B*: 5.2 |
| Burnup | *E*: 4.4 | *A*: 4.1 | *B*: 4.6 | *C*: 4.5 | *D*: 5.1 |
| Spacegas | *D*: 4.7 | *E*: 4.4 | *A*: 4.3 | *B*: 4.6 | *C*: 5.5 |
| Gojet | *B*: 4.7 | *C*: 5.2 | *D*: 5.3 | *E*: 4.7 | *A*: 4.9 |

The engines are denoted by the letters *A*, *B*, *C*, *D* and *E* and are arranged in a Latin square design. Analyse the data to investigate the fuel, injection rate and engine effects.

**4** In an experiment to determine the factors affecting the time taken for parrots to learn how to say 'gotcha', the following data was obtained:

| Time taken (weeks) | African grey parrots | | | Blue-fronted Amazon parrots | | |
|---|---|---|---|---|---|---|
| | Spoken to | | | Spoken to | | |
| Size of owner's family | every day | every 2 days | never | every day | every 2 days | never |
| Small | 5.6 | 5.6 | 5.5 | 5.2 | 6.9 | 6.2 |
| | 5.3 | 5.2 | 5.3 | 5.3 | 5.5 | 6.1 |
| | 5.7 | 6.2 | 6.7 | 5.0 | 7.0 | 6.6 |
| | 5.5 | 6.2 | 4.9 | 5.8 | 5.4 | 6.4 |
| Medium | 5.6 | 5.8 | 5.7 | 4.9 | 5.9 | 5.3 |
| | 6.0 | 5.4 | 5.9 | 6.3 | 5.8 | 4.9 |
| | 6.0 | 4.6 | 5.9 | 6.2 | 5.3 | 6.0 |
| | 6.3 | 5.2 | 5.7 | 5.6 | 5.8 | 4.5 |
| Large | 6.2 | 6.1 | 5.7 | 6.3 | 5.9 | 6.1 |
| | 6.4 | 6.5 | 6.9 | 5.5 | 6.1 | 5.6 |
| | 7.0 | 6.0 | 6.1 | 6.4 | 6.2 | 6.6 |
| | 6.3 | 6.1 | 5.4 | 6.4 | 5.7 | 6.2 |

Analyse the data. (The sum of ($x^2$) is 2487.96.)

**5** Three trawlers were fitted with underwater loudspeaker systems to investigate the effect of music on the size of the catch. In twelve sailings each boat transmitted three types of music through the amplifiers although for technical reasons these could not be the same types for every boat. The numbers of fish caught were as follows:

| Catch (hundreds) | Trawler | | | | | | | | |
|---|---|---|---|---|---|---|---|---|---|
| | Saucy Sue | | | Luscious Lucy | | | Naughty Nita | | |
| Type of music | Jazz | Opera | Folk | Popular | Orchestral | Country | Choral | Piano | Military |
| Sailings | 26 | 25 | 25 | 23 | 24 | 23 | 26 | 23 | 27 |
| | 25 | 26 | 29 | 24 | 25 | 24 | 30 | 25 | 25 |
| | 24 | 26 | 29 | 22 | 22 | 25 | 24 | 26 | 23 |
| | 26 | 28 | 23 | 27 | 24 | 22 | 25 | 26 | 24 |

Analyse the data as a nested design with fixed effects. (The sum of $x^2$ is 22 685.)

# 9 Sampling and sample surveys

There are subject areas in which the methods of planning an experiment described in the last chapter are largely irrelevant. It was assumed there that random samples and an accurate measuring instrument were always readily available. However, in marketing research, for instance, a sample has to be selected of people who exhibit the type of consumer behaviour being studied, such as buying washing machines. Furthermore a questionnaire or other data collection mechanism must be developed to measure that behaviour in some sense. We define a **questionnaire** to be any prepared form or sequence of questions designed to elicit information. It can be completed by the respondents themselves or by **field workers** who communicate with or simply observe them. There is also supporting documentation like covering letters for postal surveys or briefings for interviewers to be produced. The operation of examining a sample of respondents in this way is called a **sample survey**.

This chapter is in four sections. The first consists of an overview of the running of a survey, the next deals with sample selection techniques, the third with questionnaire design and the last one with the analysis of responses. The analysis phase of a sample survey tends to be relatively straightforward and rarely necessitates anything more elaborate than $t$ and $\chi^2$ tests. For this reason most of the work presented here is independent of the rest of the book.

## Running a survey

The important stages of planning and executing a data collection exercise were discussed in Chapter 1. We outline them here and include other considerations especially appropriate to sample surveys. The process of running such a survey is illustrated in Fig. 9.1.

The initial stage of any statistical study is to understand and define the research problem. Before planning a survey it is extremely important to identify the population of interest, called the **target population**, and specify the hypotheses to be investigated. These considerations influence the nature of the variables we attempt to measure as well as the methods of analysis which will ultimately be applied. They particularly affect the decision of whether to gather fresh information or to utilise existing data collected for some other purpose. The advantages and disadvantages of primary and secondary data were mentioned in Chapter 1. We proceed here on the assumption that primary data is to be collected although the analysis section of this chapter also relates to secondary data.

In a sample survey primary data consists of observations of or communications with members of the target population. Communication can be by **structured** or **unstructured** interviews by field workers, possibly over the telephone, or by respondents completing a questionnaire which is distributed by hand or by post. The mode by which a questionnaire is administered greatly affects its design and even the sampling methodology employed for its distribution. Questionnaires for self-completion by

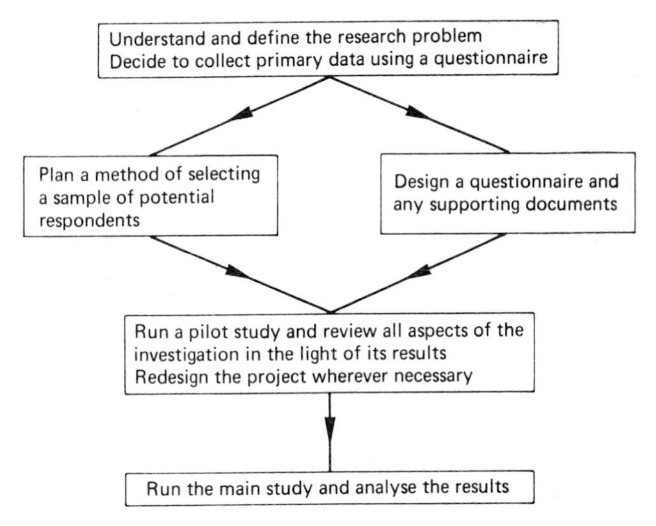


**Fig. 9.1** Running a Sample Survey

respondents are obviously quite different from those designed for use by a field worker interviewing a respondent.

Once the variables to be measured or, at least, the broad area of inquiry specified, and the method of questioning have been decided upon, the parallel problems of sample selection and questionnaire design can be addressed. These activities are covered in detail later in the chapter, so let us suppose now that we have a questionnaire and a methodology for choosing potential respondents. Before using a questionnaire it is highly advisable to run a **pilot survey**. This is a miniature version of the real thing which tests all aspects of the proposed operation. The results are treated as feedback with modifications being made wherever necessary to the questionnaire layout, the individual questions and any supporting documents like field staff briefings and covering letters. The researcher must be prepared to judge harshly on whether the specific objective of each question has been achieved and whether the questionnaire as a whole is capable of delivering the information required. It is not just the questionnaire that needs calibrating at this point: the briefing and performance of field workers, the **response rate** of a postal survey and other related factors should also be reviewed.

In a well-organised sample survey the distribution of the final draft of the questionnaire is something of an anticlimax. Of course with hindsight the researcher is always able to suggest improvements to the survey design. The most evil gremlin in the author's experience of many years of consultancy work in this area is the low response rate of postal surveys. Typically only 30% of questionnaires sent out are returned despite the inclusion of a covering letter, a stamped addressed envelope and, to companies, the promise of a summary of the results for their information. Chasing up non-responses by letter or telephone is not satisfactory as it distorts the replies. Respondents want to dispose of the questions as quickly as possible and give superficial and inaccurate answers in order to do so. In some government surveys the completion of a questionnaire is mandatory on the respondent, for instance in applying for a passport. Governments also carry out a census from time to time, which is not strictly speaking a sample survey as everyone is obliged to respond. Incidentally, the published conclusions from these regular censuses constitute a valuable source of secondary data for validating the efficacy of particular questions in sample surveys.

The analysis of questionnaire data is the subject of the last section of this chapter. It is the point in the study where some conclusions are reached but more research problems are raised. It is the mark of a good analysis to identify objectives which have not been achieved and to reveal whole new areas for future investigations. The process of scientific inquiry never ends but proceeds from one level of knowledge to the next.

## Selecting a sample

A **sampling frame** is a list of the members of the target population. In many studies it does not exist or is unavailable to the researcher. There may be other lists which do not constitute proper sampling frames for various reasons. For example a telephone directory is not a sampling frame of householders as it does not contain ex-directory numbers or homes without a telephone. Even reliable sampling frames like electoral registers become out of date remarkably quickly. In spite of these problems it is always advisable to attempt the acquisition of as accurate a sampling frame as possible. It enables a representative sample to be chosen and consequently reduces the risk of bias in the final conclusions.

We now consider some methods of selecting a sample. They are presented in decreasing order of statistical desirability. The first few require a sampling frame while the remainder can, and often do, give correspondingly poor results.

*Random sampling*
In this method every single member of the target population has exactly the same probability of being included in the sample. The random sample this produces is the basis for the theory behind all the standard statistical tests and procedures. The items in the sampling frame are numbered and tables of random numbers consulted to select them. Suitable tables can be found in books like reference (66) of the Bibliography. Alternatively, random numbers can be generated by a computer or the digits from lists of telephone numbers and the like, provided they are truly random, can be used.

*Systematic sampling*
This is a variant of random sampling whereby only the first item in the sample is chosen at random. Subsequent items are selected from the sampling frame at regular intervals from that point onwards. For instance, suppose the sampling frame consists of 2000 subjects and we want a sample of 500. A random number between 1 and 4 provides us with the first item. This could be obtained by rolling a die until such a number occurs. Selecting every fourth item thereafter through the sampling frame gives a sample of the required size. It can be argued that if the sampling frame lists the subjects in some logical sequence, like alphabetical order, then this method produces a representative sample across that classification.

*Stratified sampling*
There is often an attribute, like age, which divides the target population into subsets called **strata**. Stratified sampling is the selection of a random sample from each stratum. If the size of the sample is proportional to the number of items in the stratum, then the method is **proportionate stratified sampling**. Sometimes the survey is concerned with a group of specific strata and larger samples are taken from them than their size warrants. This is called **disproportionate stratified sampling**. For example, a survey on geriatric care might take a larger sample of elderly respondents than their occurrence in the population as a whole suggests. That particular sample could then be subjected to a more detailed analysis.

*Cluster sampling*
Many populations fall into natural and representative groupings like families or geographical localities. One or more of these clusters can be selected at random and either the whole group used as a sample, called **one-stage cluster sampling**, or random samples chosen from each one, called **two-stage cluster sampling**. It has the cost-beneficial consequence that field workers need visit only a relatively small geographical region rather than tour an entire area interviewing respondents scattered far and wide.

*Other methods*
The procedures above presuppose the availability of a sampling frame even if only within the clusters of cluster sampling. We now consider techniques which do not require such a list but are correspondingly less desirable as they give unrepresentative and self-selecting samples. **Quota sampling** is popular in marketing research and opinion poll surveys. Each field worker is given a quota of respondents like '20 middle-aged men carrying briefcases'. Clearly the field workers' perceptions of the description itself and of how it should be applied affect the extent to which the sample can be representative. Respondents who appear approachable or otherwise attractive will be selected. Respondents who see the survey in progress and volunteer information will likewise be accepted. The sample is therefore not representative and certainly not random.

In **judgement sampling** subjects are selected because they are judged to be indicative or typical of the population as a whole. For instance in the period before an election voters in a marginal constituency are canvassed in the hope that they act as some sort of political barometer. Similarly in economic surveys 'captains of industry' are consulted about the future prospects for commerce and the economy.

**Accidental sampling** is the least statistically desirable method of all. A sample is formed of those subjects who present themselves as respondents. Radio phone-in programmes are a good example of this although basing ideas about public opinion on the letters received by women's magazines comes a close second. Large organisations sometimes have forms on which customers can register suggestions or complaints, and while they may prove valuable for gathering certain types of information they cannot be considered as providing a random or representative sample. This kind of sample is completely self-selecting and may be likened to eating spaghetti with a fork, only the longer strands are picked up.

*Sample size*
It is difficult to give general rules about sample size. As discussed in Chapter 1, sample size depends on population variability. If every single item in the population were exactly the same then a sample of just one would be sufficient. Some calculations can be performed with the normal or $t$ distributions. For example, suppose a pilot survey indicates that the standard deviation of a certain variable is around 25. Suppose further that we want to be 95% certain of estimating the population mean to within 3 units. From the work contained in the Crisps Example of Chapter 3, we want

$$\frac{1.96\,s}{\sqrt{n}} = 3 \qquad (9.1)$$

where $s$ is the standard deviation and $n$ is the sample size. It is assumed here that $n$ is so large that the normal distribution can be used.

Substituting 25 for $s$ and solving this equation for $n$ we find that it is 266.78. It would therefore be wise to take a sample of 267 or more items.

The above argument is not very helpful in practice. A sample survey is usually concerned with more than one variable and the researcher cannot quantify how accurately the population mean is to be estimated. There are also many other objectives of a survey in addition to the estimation of simple population parameters. The correlation between variables is often of importance.

## Designing a questionnaire

Nowhere does statistics appear more as an art than in the design of questionnaires. The topic is a blend of data collection and applied psychology which requires patience and a level of self-criticism bordering on the obsessive. Questions must be polished and repolished until they behave in precisely the manner intended. There are virtually no second chances with sample surveys and respondents cannot usually be approached twice. It is therefore necessary to be fully aware of the variables to be measured and the level of measurement which can be attained for them. This applies particularly to attitudes or opinions which are being gathered from the sample.

The analysis envisaged and the format of the conclusions should be borne in mind throughout the design phase. Only those questions which will serve a meaningful purpose in that phase should be included. We study first the mechanics of asking questions and examine more general aspects of the questionnaire afterwards.

*Question structure*
The nature of a question is determined by the type of variable it has to measure. Nominal, ordinal and metric variables were introduced in Chapter 1 and it is assumed here that their characteristics are understood. Nominal data is very common in questionnaires and is relatively easy to elicit. A **dichotomous** question gives the respondent a choice of two answers, for example 'Do you own an umbrella?' has the possible responses 'Yes' or 'No'. The provision of a 'Don't know' reply to such questions should always be considered.

A **multichotomous** or **cafeteria** question contains a menu of possible responses. For instance the question 'Which of the following sports do you like?' might have a list of sports for the respondent to tick, underline or circle. One of the categories can be 'Other' or 'None of the above' or can be a dotted line for a reply to be written in an open-ended way. This type of question often allows **multiple responses** in that more than one option can be selected from the menu. The researcher should have decided how the data from such questions is to be analysed before including them. In the above example, the rating of different sports on a scale from, say, 1 to 5 or the ranking of the sports in decreasing order of preference would convey more information.

Most dichotomous and multichotomous questions are **structured** as the respondent opts for one or more possible answers chosen from a list. In an **unstructured** or **open** question respondents speak or write answers in their own words. This is coded by the field worker or researcher on the spot or afterwards using a **coding frame**. This essentially classifies the response on a multichotomous scheme which would either be too technical to include on the questionnaire or simply too big. In a medical questionnaire a patient will not understand the terminology used but a trained field worker, like a nurse, can make sensible classifications of clinical conditions. Similarly in asking the question 'In which country were you born?' it is better to let respondents write it down rather than present them with a menu of all the countries of the world.

Ordinal data is usually collected using multichotomous questions, for instance, 'Rank these composers in order of preference' followed by a menu containing Beethoven, Bach, Mozart and so on. Attitudes and opinions are measured as ordinal variables and respondents indicate a level of agreement or disagreement by ticking or circling a number along a scale. Questions to gauge states of mind are the most difficult to devise. A lot of research has been carried out on how best to validate such questions. It is common practice in marketing surveys for panels of experts to grade questions like 'Flopsy soap is kind to hands' and 'Flopsy soap is smooth on your skin' by the variety of responses they expect to get from consumers on an agreement/disagreement scale. The questions which are thought to be most controversial are included in the questionnaire along with a sprinkling of the others. In this way not only are opinions canvassed but the very parameters by which those opinions are believed to be formed, for instance 'kindness to hands' or 'smoothness on the skin', are assessed for effectiveness. In some areas these parameters are changing and are the subject of fashion, so this continual validating and recalibrating of the measuring instrument, the questionnaire, is essential. Wherever possible the behaviour rather than the thoughts of the respondent should be measured. The true attitude of a person is betrayed by what they do and not necessarily by what they say.

It was stated above that nominal data is the easiest type to measure by means of a questionnaire. All the respondent has to do is select an option from a menu of possibilities. Although plenty of damage can be caused by such a procedure it is more reliable than obtaining opinions, which are ordinal in nature, with a scale of agreement/disagreement values. When the level of measurement being aspired to is metric, then the situation can become difficult for other reasons. If the data is to be recorded and processed in its raw form, then an open-ended question like 'How old are you?' can be asked. Provided the respondent has accurate information available and the desire to answer the question properly, then all is well. However, as most of the data collected by questionnaires is nominal, correlating such variables with others can be awkward. The reader is referred to the Drinks Example of Chapter 4 to appreciate this point. If a metric variable is going to be classified into groups in the subsequent analysis anyway, then it is sensible to have the respondent perform the classification in the question itself. Besides making the question quicker to answer it is less sensitive for subjects to identify a range of, say, salary levels, than have to quote their exact earnings.

*Sequencing the questions*

It is usually advisable to put the questions into a logical sequence which 'tells a story' or otherwise develops a theme. This helps the respondent's ability to empathise with the objectives of the study and to recall past behaviour and attitudes accurately. We are essentially asking the respondent to relive any relevant experiences and take them through the situation step by step. Questions can be classified for the purpose of sequencing:

1 **Status questions** — facts like name, address, sex and age.
2 **Past or current behaviour questions** — actions which have become facts within the respondent's mind. They are or once were in his or her conscious experience.
3 **Awareness and level of knowledge questions** — contents of the respondent's mind which may never have been brought into the conscious, let alone articulated. These questions require more thought than previous ones. Recall can be **aided** by showing the respondent pictures and the like or **unaided** with a 'Can't remember' category for the response.
4 **Attitudes, opinions and motivation questions** — subconscious feelings which are difficult to quantify. These questions are difficult to compile and require careful validation in a pilot study before being used.

5 **Future intentions and behaviour questions**—the least reliable type of questions as they involve hypothetical situations being presented to the respondent who answers at random or in an attempt to impress. These issues are dealt with later.

Opinions differ among practitioners about the 'best' sequence for the various types of questions. The first two types are relatively easy to answer and there is an argument for putting them at the beginning of the questionnaire. They relax the respondent and acclimatise him or her to the general style of questioning. However, in an interview by a field worker it may be more natural to begin by talking about the subject matter. The behavioural questions then come first with the status ones at the end.

By judicious sequencing of carefully worded questions responses can be biassed in all sorts of ways. For example, consider the following two sequences leading up to the same question.

*Question 1* 'Should society provide doctors with every possible means of saving human life?'

*Question 2* 'Would you knowingly obstruct a doctor as he tries to save a life?'

*Question 3* 'Should a doctor be allowed to remove the kidneys from a person killed in a road accident in order to perform a transplant operation?'

and the second sequence:

*Question 1* 'Did you know that there are at least three different clinical definitions of what is meant by "death"?'

*Question 2* 'Do you think doctors are more likely to make a mistake or less likely to make a mistake when they are at the scene of a road accident?'

*Question 3* 'Should a doctor be allowed to remove the kidneys from a person killed in a road accident in order to perform a transplant operation?'

The two sequences clearly tell a different story, and lead the respondent in different directions. Unfortunately any controversial topic, by definition of the word controversial, will have associated with it several threads of argument pointing to various conclusions. It is hard to be impartial when framing questions but care should be taken to avoid a seductive line of questioning which produces a bias in the mind of the respondent.

*Question wording*

Words open doors and conjure up images in a respondent's thought processes. A well-phrased question opens relevant doors and conjures up relevant images. Both field worker and respondent are helped if the questions are as concise as possible without being ambiguous. Besides creating a favourable impression short questions elicit more accurate answers. Halfway through reading a long question the respondent tends to anticipate the remainder of its wording. Worse still, his or her reply to this incorrect version of the question is decided upon before the end of the question is reached. The case for being aware of subconscious connotations being triggered and for brevity extends to the whole questionnaire as well as to the wording of individual questions. The mere appearance in the post of a five-page document can deter even the most cooperative subject.

Simplicity is another consideration allied to brevity. The questions and any instructions, either to the field worker or the respondent, must be written in a form of language and use vocabulary that is easily understandable. Particular care is necessary with supplementary questions like 'If you replied "Yes" to the last question, how often do you do it?' and 'If you replied "No" to at least one of the last three questions go to question 96'. Although it is obvious that questions like 'Have you stopped beating your wife?' are impossible to answer without admitting incriminating behaviour, there are more subtle traps for the questioner. As an experiment, the next time you ask someone a question and would like a 'yes' answer, nod your head slowly as you speak the question. Similarly, a 'no' answer is encouraged by slowly shaking your head. The fact that this kind of message from interviewer to interviewee works must be taken into account in question wordings.

Respondents have to be able to answer our questions. Firstly the information required must exist and secondly be within their immediate recall or power to access. Ambiguous words like 'nice' and 'good' are to be shunned as they have different meanings for different people. Having made this point there are occasions when the very perception of what is 'good', with all its ambiguities, needs to be measured. The act of classifying something as 'good' by a respondent can be revealing in an area of study like consumer behaviour.

*Presentation*

The appearance of the questionnaire as a whole creates an important first impression on the respondent. The degree of professionalism which has been brought to bear in the preparation of all the documentation including field workers' briefing notes and covering letters is now on display. In some cases the respondent has a deeper knowledge of the subject matter than the researcher. For instance a statistician uneducated in medical affairs may design a questionnaire for surgeons about the materials and instruments they prefer to use in operations. In all sample surveys the researcher is approaching the respondent, via the questionnaire and other documents, in order to request information. Artwork which is pleasing to the eye and professionally produced prompt cards or menus for interviewers to show subjects facilitate this flow of information. For postal surveys a covering letter explaining the purpose of the study helps in gaining the support of the respondent. Remember to include a stamped addressed envelope for the reply! Sometimes non-responses are followed up by post or by telephone but the quality of the data so obtained is not always very high. Given that a subject did not 'take the bait' in the first instance it is doubtful whether they will give the questionnaire serious consideration thereafter.

*Data-processing aspects*

The vast majority of sample surveys are analysed by computer. Hence every response on every questionnaire has to be keyed into the system being used. Each questionnaire constitutes a **record** with the response to each question being a **field**. Each response is coded as a number with zeroes being crossed through to distinguish them from the letter O. A set of records representing several questionnaires is called a **data file**. For example, four questionnaires, each with eight responses on them, might appear as a file of four records like this:

5,5,2,3,12,2,9,14
2,2,4,3,1Ø,4,7,22
Ø,Ø,1,1,42,5,3,25
Ø,6,8,9,81,4,Ø,15

In many computer systems fields are separated by commas as above, although sometimes spaces are used.

Figure 9.2 shows some examples of how responses to various types of questions can be coded. It is obviously advisable to involve the operators who will key the data into the computer in the design of the questionnaire. The coding arrangements should be agreed with them before the document is printed, particularly the positioning of the field numbers giving the order in which the responses are to be typed to form a record.

Although different methods of recording responses are illustrated in Fig. 9.2, it is less confusing if boxes or circled numbers are used consistently throughout. Many questionnaires do not return accurate information because the instructions to the respondent are inadequate or unnecessarily complicated. This applies especially to cafeteria questions in which the requirement of multiple responses or a single response

Please indicate your age by ticking the appropriate box:

| less than 20 years | ☐ 1 | between 20 and 60 years | ☐ 2 | over 60 years | ☐ 3 |
|---|---|---|---|---|---|

(Field 7)

Circle the figure 1 against any of the foods in this list that you have dreams about:

| | | | | | | For office use<br>Field |
|---|---|---|---|---|---|---|
| Marshmallows | 1 | Chocolate | 1 | Syrup | 1 | 6, 7, 8 |
| Apple pie | 1 | Ice cream | 1 | Porridge | 1 | 9, 10, 11 |
| Other (please specify) ........................................ | | | | | | 12 |

Circle *one* number in each row to indicate how often you do the following activities:

| | Often | Fairly often | Hardly ever | Never | |
|---|---|---|---|---|---|
| Sleep in the bath | 1 | 2 | 3 | 4 | (22) |
| Eat at midnight | 1 | 2 | 3 | 4 | (23) |
| Buy bananas | 1 | 2 | 3 | 4 | (24) |

Put the following television programmes in order of preference using 1 for 'best', 2 for 'second best' and 3 for 'worst':

| | Rating | |
|---|---|---|
| 'Win a Nuclear Power Station' | | (18) |
| 'Moontrip' | | (19) |
| 'Scapegoat' | | (20) |

How much do you sell 2 litre bottles of lemonade for? | £ : p | (11)

**Fig. 9.2** Possible coding arrangements for various types of questions

must be indicated. The researcher will want to distinguish between the refusal of a respondent to reply to a question and its lack of relevance to him or her. Both of these possibilities result in a blank answer unless a category like 'Don't know' or 'Not applicable' is included in the menu. A code digit, usually zero, should be reserved to indicate 'no response' with all other replies from the respondent being coded as 1, 2, .... If a coding frame is being employed to classify responses for the purpose of computer data entry, then it should be unambiguous and easy to use.

*Confidentiality*
Many surveys touch on areas which respondents might find sensitive and require assurances that confidentiality will not be breached. The researcher should always consider giving respondents anonymity in such cases. Many questionnaires ask for names and addresses when in fact the information is not really needed. These questions alert the respondent to potential identification and should be omitted where possible.

Questionnaires on sensitive issues can be prefaced by an assurance from the interviewer, or accompanied by a letter, explaining the precautions being taken against improper access to the information given. At a more sinister level, a delicate question can be inserted into a sequence of innocuous ones and so be answered when the respondent's defences are down. It is also possible to phrase questions to disguise their true purpose and yet still measure the relevant variable. For example, rather than ask 'Did you murder your mother-in-law?' the respondent can be reassured with the inquiry 'Lots of people are murdering their mothers-in-law these days. Are you one of them?' The question can be put into a hypothetical form, 'If your mother-in-law became difficult and you lost your temper, do you think you could kill her?' The third-person mode can be used, 'If a man became so provoked by his mother-in-law's behaviour that he killed her, would you blame him?' There are other versions of this theme and they take us well into the domain of applied psychology.

*Cosmetic responses*
These are biassed responses which favour the respondent's status or performance in order to create a good impression. Some questionnaires set out to assess these 'perceived self-portraits' but in most studies cosmetic responses are a nuisance. They even occur when people complete a questionnaire on behalf of a company and declare excessive turnover figures or inflated profits.

**Check questions** can help measure the bias introduced by respondents or the extent to which they have lost interest in the questionnaire and are answering at random. Such questions either make the same inquiry in a different way somewhere else in the questionnaire or merely ask questions with known answers. Wherever possible behavioural questions rather than attitude ones should be used. The degree of bias from cosmetic responses is affected by the respondent's perception of the nature and purpose of the survey. When asked about last year's earnings a young man's reply to an attractive female interviewer may well be different from his response on the taxman's form.

## Analysing the data

Having designed and piloted the questionnaire, we run the survey and hopefully collect some data. The estimates of population parameters and the hypotheses to be tested have all been decided in advance as they dictate which questions are to be asked. The analysis phase is therefore merely the filling in of prepared tables with the appropriate frequencies, averages and so on.

Before reviewing analysis techniques it is sobering to consider the reliability of questionnaire data as stored in a computerised system. Figure 9.3 shows the errors

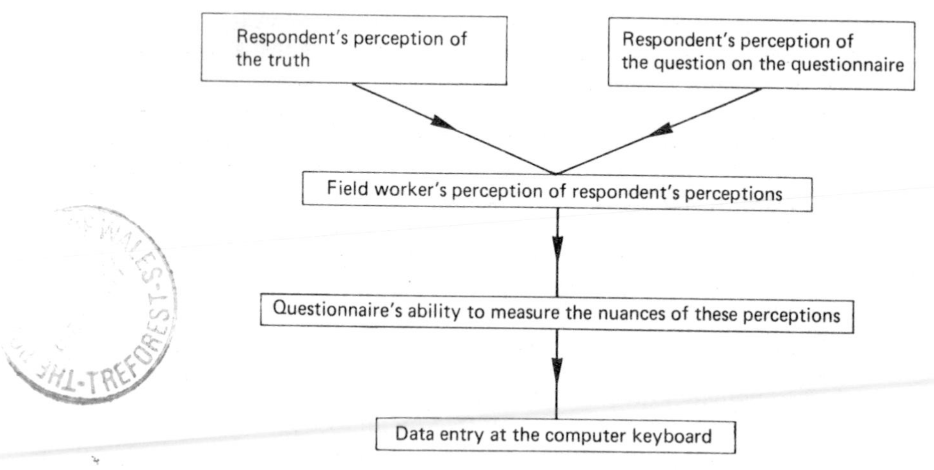




**Fig. 9.3** Sources of errors in Sample Survey data

which can creep into it en route from being the respondent's version of the truth to being digits in a computer.

Each box in Fig. 9.3 represents a possible source of error, although for a self-completion questionnaire the 'field worker' box is not present. The existence of an interviewer between the respondent and the questionnaire undoubtedly reduces the accuracy of certain types of data. On the other hand, a field worker is indispensable for technical surveys or where time is important, as in an opinion poll.

It is usual to begin the analysis of questionnaire data by calculating the frequency of each response for each question. This produces a **one-way tabulation** containing a frequency for each possible reply to a question. The information can be displayed diagrammatically with bar charts, pie charts and so forth or described numerically with averages, variances and other measures of position and spread. Where appropriate, confidence intervals for metric variables or population proportions of attribute variables can be quoted. All these methods were covered in Chapters 2 and 3 of this book.

Contingency tables show the frequencies of responses to pairs of questions as in the Drinks Example of Chapter 4. In survey work they are often referred to as **cross tabulations** or **two-way tabulations.** The row and column totals of a two-way table are the one-way tables of frequencies for each of the two questions concerned. It is thus common to present such tabulations with rows and columns totalled.

**Many-way tabulations** correlate the responses to more than two questions. They can be arranged as collections of two-way tables. For instance, if a questionnaire classifies respondents by sex, age and food preference, then there will be a two-way tabulation of food preference by age for men and another two-way tabulation for women. The format of the data in the Straight Line Example of Chapter 8 could be used to convey this three-factor structure, although the analysis would be carried out on the individual two-way arrays.

Ordinal data and multiple response attribute data are the most difficult types to analyse. Respondents' first choices in ranking items can be tested for significance with the $\chi^2$ distribution as in the Newspaper Example of Chapter 3. Care should be taken in examining second choices. The expected frequencies must be calculated allowing for the fact that respondents cannot rank the same item as being first and second. Hence the expected frequency an attribute has of being the second choice depends on the number of respondents who did *not* opt for it as their first choice. This idea is developed in Question 5 of the Exercises below. Alternatively each option can be given a score according to the rank it receives from the respondent. The total scores for the whole survey then serve to rank the options.

Apart from the $\chi^2$ and $t$ distributions, correlation, regression and analysis of variance methods can all be utilised on sample survey data. A questionnaire contains measurements of many variables and so the multivariate techniques of Chapter 5 are also extensively used. Cluster analysis is particularly appropriate and helps us to identify groups of respondents with similar attitudes or other characteristics.

A survey should not be analysed in isolation but validated wherever possible against external secondary data. For instance if there have been no quotas placed on the ages of the respondents, then the age ranges obtained should agree with those of a government census. Such validation of sample selection procedures and field worker impartiality is extremely valuable.

Finally a survey should always be viewed as an intermediate stage in the information-gathering process. Of necessity the design of the study follows on from the state of knowledge the researcher has at the start of the operation. Having analysed the results, he or she should be prepared to proceed to a further investigation involving more penetrating questions. This spirit of scientific inquiry applies equally to routine marketing research surveys in which the performance of the questions can be regularly assessed and enhanced by careful interpretation of the data. In these ongoing areas of investigation the real research is about the efficiency of the measuring instrument, the questionnaire, and its application. Running this year's study should involve more than bringing out last year's questionnaire and blowing the dust off it.

## Summary

**1** The important stages of a sample survey are understanding the research problem, selecting the sample, designing a questionnaire, running a pilot study, running the main study and analysing the results. It is important to feed the information gained from the pilot exercise back into the design process before the main study takes place.

**2** A sampling frame is a list of all the members of the target population. If it exists then random sampling or systematic sampling can be used. Stratified sampling views the target population as being made up of separate strata according to some attribute like age. Cluster sampling views the target population as being made up of groups of items, each one being fully representative of the whole. There are numerous other methods of selecting a sample like quota sampling, judgement sampling and, least representative of all, accidental sampling.

**3** Determining the sample size to employ is difficult. A questionnaire contains many variables and basing the decision on any one is unsatisfactory. There are no hard and fast rules.

**4** Questions on a questionnaire can be dichotomous, multichotomous or open. They can be classified as status, behavioural, awareness, attitudinal or future intentions questions. Their sequence in a questionnaire affects the respondent's recall ability and general impression of the survey. This applies equally to question wording, which should be simple and brief. The information required must be close at hand for the respondent and within his or her ability to give. The physical appearance of the questionnaire and any supporting documents is important. Any prompt cards or menus the field workers use should be of an acceptable standard. Each response has to be coded for computerisation. There are ways of minimising the sensitivity felt by respondents to giving certain types of information. In fact psychology has a lot to offer in helping researchers design questionnaires. This aspect of the situation extends to dealing with the problem of cosmetic responses.

**5** The number of places where errors creep into survey data must be borne in mind by the analyst. Tabulations of the frequencies can be produced along with graphs, charts and numerical measures. Much of the work in other chapters of this book is relevant to survey analysis, although most studies do not call for sophisticated techniques. The results should be checked with other surveys in order to validate the methodology. As in other branches of science, an analysis is simply an intermediate step and further studies and experiments are suggested by it to take our knowledge of the underlying research problem forward.

## Further reading

A range of books on questionnaire design and sample surveys is given as a separate section in the Bibliography at the end of the book. Some are more mathematical than others, but (46) and (51) are written in simple terms with good examples. They contain work on attitude measurement and its problems and classifying respondents by socio-economic groups. References (49) and (50) have been included as they allow a researcher to identify sources of secondary data for direct use or for validating a survey. Government departments are excellent start points for locating such sources. Some general statistics texts, like (9), deal with sample-selection methodology and sample-size implications.

It is unfortunate that sound questionnaire construction and good survey practice cannot be effectively learnt without 'hands-on' experience with a skilled practitioner. As stated earlier, this is the area in which statistics most appears to be an art rather than a science.

## Exercises

**1** Identify those features of a problem which lead a researcher to adopt a sample survey as an appropriate research method.
**2** You are chief statistician on board a spaceship mission to a hitherto uncharted galaxy. It is known that intelligent life exists on the planet being approached and your commander asks you to prepare a questionnaire to be used on the creatures found there. Write a report explaining why this is not possible until you have met the aliens, even though they speak English.
**3** Design a questionnaire to determine the flavour, colour, constituency, packaging and price a proposed new brand of sweets should have to appeal to children aged between 10 and 14 years.
**4** The numbers of respondents to a survey who preferred different brands of cornflakes were classified according to the country they live in:

| Number of respondents | Preferred brand of cornflakes | | |
|---|---|---|---|
| | Crunchers | Soggies | Popples |
| England | 25 | 30 | 25 |
| Scotland | 10 | 15 | 32 |
| Wales | 15 | 15 | 33 |

(i) The column totals represent all those respondents who favoured each brand. Do they indicate that the three types of cornflakes differ significantly in their popularity? (ii) The manufacturer of Popples claims that this product commands at least 50% of the market. Do the column totals contradict this claim? (iii) Determine whether there is an association between the brand of cornflakes preferred and the region in which a respondent lives.

**5** The rankings given to three brands of cat food by 168 respondents were:

| Number of respondents | Rank | | |
|---|---|---|---|
| | 1st | 2nd | 3rd |
| Catto | 43 | 24 | 101 |
| Philcat | 73 | 58 | 37 |
| Fatpuss | 52 | 86 | 30 |

(i) Are respondents discriminating between the brands in making their first choice? In other words, test the first column of frequencies to determine whether they are choosing at random between the three options. (ii) If respondents are making their second choice at random, then we expect the number choosing Catto to be half of those who did not make it their first choice. This is because they have chosen Philcat or Fatpuss and now have a choice of two foods for second place of which Catto is one. Hence the expected frequency for Catto as second choice is $(168-43)/2$ which is 62.5. Calculate the corresponding expected frequencies for Philcat and Fatpuss. (iii) Test the second column of frequencies to decide whether second choices are being made at random

# 10 Quality control

Industrial consumers and producers are concerned about the quality level of raw materials, component parts and finished products. The manufacturing process converts one type of commodity into another and if the materials bought from a supplier are sub-standard, then the final article will also be sub-standard. Statistical quality control relates to two kinds of situations: the monitoring of a production process as it progresses and the decision of whether to accept a batch of components as a consumer.

There are four sections in this chapter. The variable being controlled in the monitoring of a production process is often assumed to be distributed normally. The first section examines how a normal distribution is fitted to a sample of data and assessed for validity. Life testing of materials and components is an important area of quality control, and the second section deals with reliability and the Weibull distribution as applied to it. There follows a treatment of control charts which help a production manager to monitor a manufacturing process. The chapter ends with an example of acceptance sampling in which the decision to accept a batch of items is based on the information contained in a sample drawn from it.

## Fitting a normal distribution

The Rice Example of Chapter 3 demonstrated the use of the normal distribution and the associated tables on pages 183 and 184. The $\chi^2$ distribution was also introduced in that chapter to test the goodness of fit of a set of expected frequencies to a set of observed ones. We bring those ideas together in the example below in determining whether a variable being monitored for quality-control purposes is normally distributed.

### The concrete example

A company manufactures and sells ready-mixed concrete. The quality of the concrete is measured by the force a cubic metre can withstand before disintegrating. The pressures at which a sample of 472 cubes broke were grouped into classes as follows:

| Strength (N/mm$^2$) | Less than 25 | $25^+$–30 | $30^+$–35 | $35^+$–40 | $40^+$–45 | More than 45 |
|---|---|---|---|---|---|---|
| Number of cubes | 54 | 70 | 117 | 119 | 72 | 40 |

*Question 1* Fit a normal distribution to the strengths of the cubes.

*Solution* A normal curve is characterised by the mean $\mu$ and standard deviation $\sigma$ of the population. In the Rice Example they were estimated from $\bar{x}$ and $s$ calculated for a large sample, and although the raw data has been grouped here we could adopt the method of the Dandelions Example of Chapter 2. This provides a short-cut technique if required. However, rather than approximate data values by the mid-points of the classes they belong to, it is better to utilise cumulative frequencies which can be read accurately from the given table.

Table 10.1 shows the working involved in determining $z$ scores for the $x$ values in the data. The first two columns follow from the information supplied. The third column is obtained by dividing the entry in the second column by the sample size, 472. The $z$ scores are found from the percentage points table on page 184 and represent what the $x$ values in the first column should scale to if the data is in fact described by the normal distribution. In each case the $z$ score is such that the area to its right is the probability in the third column. The symmetry of the normal curve is utilised here and the reader is referred to Chapter 3 for revision of the methodology if necessary. Alternatively the computer program on page 185 can be used to evaluate the $z$ scores. By either approach, we derive a set of theoretical $z$ values to be compared with the $x$ values in the first column.

**Table 10.1** $z$ scores for the Concrete Example

| | Strength, $x$ (N/mm$^2$) | Number of cubes with greater strength | Probability of cube having greater strength | $z$ score of the probability from standard normal percentage points table |
|---|---|---|---|---|
| | 25 | 418 | 0.89 | −1.23 |
| | 30 | 348 | 0.74 | −0.64 |
| | 35 | 231 | 0.49 | 0.03 |
| | 40 | 112 | 0.24 | 0.71 |
| | 45 | 40 | 0.08 | 1.41 |
| Total | 175 | — | — | 0.28 |
| Average | 35 | — | — | 0.056 |

The relationship between $x$ and $z$ was discussed in Chapter 3, and equation (3.1) is repeated here in a slightly different form:

$$z = \left(\frac{1}{\sigma}\right)x - \left(\frac{\mu}{\sigma}\right) \tag{10.1}$$

This implies that a graph of $z$ against $x$ should be a straight line with slope $(1/\sigma)$ passing through the point $(\bar{x}, \bar{z})$. The analysis of a linear relationship between two variables is dealt with in the Cholesterol Example of Chapter 4. In applying a regression analysis here we take $x$, which is free of errors, to be the independent variable and $z$, which has been calculated from the observed frequencies, as the dependent one. Figure 10.1 is a scatter diagram of the 5 points we have numerical values for together with their average point $(35, 0.056)$. A line can be drawn by inspection or its equation calculated as in Chapter 4, while if the points do not lie in a straight line then the data values are not adequately described by the normal distribution. If they are in the shape of a letter 'U' then each $x$ value has fewer items in the sample to its right than predicted by the normal curve. Hence the $x$ distribution is not symmetric and has a mean which is to the left of centre. Conversely, if the points on the scatter diagram are curved the other way, like an upside-down letter 'U', then the mean of the parent

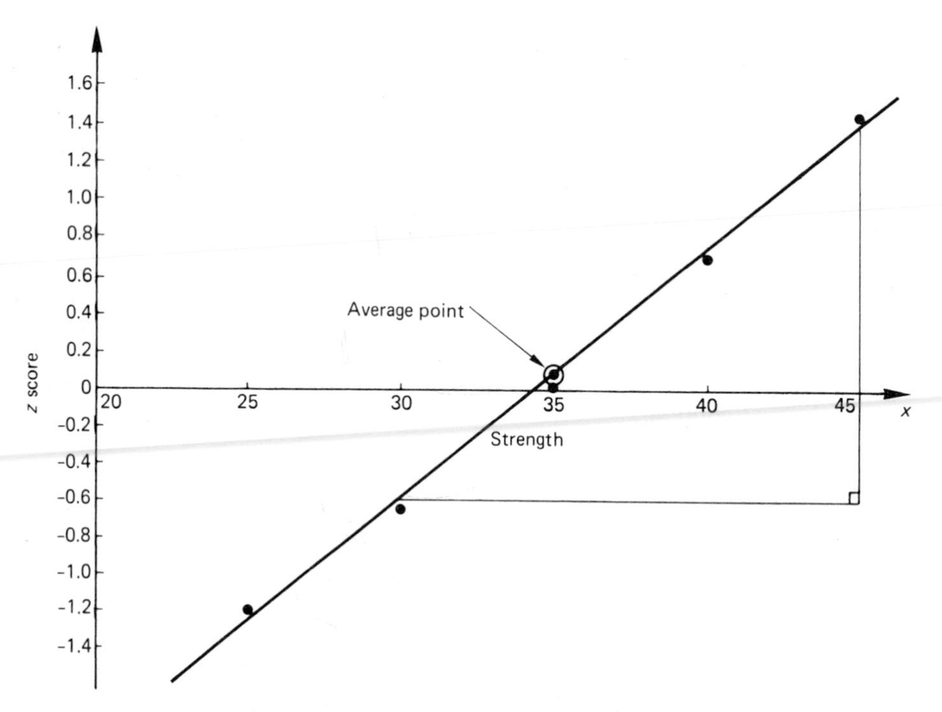


**Fig. 10.1** Regression line for the Concrete Example

distribution is to the right of centre. The extent to which a straight line explains the relationship between $x$ and $z$ can be tested using the methods of regression analysis in Chapter 4. Let us suppose here that the line in Fig. 10.1 has been drawn by eye through the average point.

We can now estimate the values of $\mu$ and $\sigma$ from the regression line. The point where the line crosses the $x$ axis provides an estimate of $\mu$. It is the $x$ value corresponding to the $z$ score zero and hence is the mean of the distribution. Having a numerical value for $\mu$ we could then substitute the averages of $x$ and $z$ into equation (10.1) to determine $\sigma$. As a more accurate alternative a right-angled triangle is shown in Fig. 10.1, and this enables the slope to be calculated. The height of the triangle, as measured on the vertical axis, is 1.96 and its base, according to the horizontal axis, is 15. The slope of the line is the height divided by the base, and from equation (10.1) is an estimate of $(1/\sigma)$. Hence $\sigma$ itself is approximately (base length/height), which is 7.65. Transposing $\mu$ to be the subject of equation (10.1) and substituting $\bar{x}$ and $\bar{z}$ for $x$ and $z$, we have:

$$\mu = \bar{x} - \sigma\bar{z} = 35 - (7.65)(0.056) = 34.57 \qquad (10.2)$$

**We have now fitted a normal distribution to the data and estimated the population mean to be 34.57 N/mm$^2$ and the population standard deviation to be 7.65 N/mm$^2$.** The above analysis can be carried out on special normal probability graph paper. The vertical axis is calibrated in a non-linear way so that the transformation from probability to $z$ score is achieved automatically as the points are plotted.

*Question 2* Test the hypothesis that the data comes from a normal population.

*Solution* The regression performed above could be tested by means of the correlation coefficient as described in Chapter 4. Table 10.2 shows a method using the $\chi^2$ distribution, which allows us to see whereabouts the normal curve agrees with the observations and where it does not. It is analogous to Tables 3.3 and 3.5 for the binomial and Poisson distributions. The third column is a $z$ score calculated with $\mu$ equal to 34.57 and $\sigma$ equal to 7.65 as found in the solution to Question 1. The probabilities in the fifth column are obtained by subtracting the entry in column 4 in the next row from the current entry in column four. This produces the area under the standard normal curve corresponding to the appropriate class. The expected frequency in the next column is the probability multiplied by the total number of cubes, 472. Finally the contributions to $\chi^2$ and their total are evaluated in accordance with equation (3.17). The number of degrees of freedom is given by equation (3.19) and in this case is the number of predicted frequencies, 6, minus the number of calculated quantities, 3. The three quantities were the total frequency, the mean and the standard deviation.

**Table 10.2** $\chi^2$ calculation for the Concrete Example

| Strengths (N/mm$^2$) | Observed frequency | $z$ score of left-hand endpoint | Area to right from normal tables | $P$(class) | Expected frequency | Contribution to $\chi^2$ |
|---|---|---|---|---|---|---|
| Less than 25 | 54 | — | 1.000 | 0.106 | 50.0 | 0.32 |
| $25^+$–30 | 70 | −1.25 | 0.894 | 0.168 | 79.3 | 1.09 |
| $30^+$–35 | 117 | −0.60 | 0.726 | 0.250 | 118.0 | 0.01 |
| $35^+$–40 | 119 | 0.06 | 0.476 | 0.237 | 111.9 | 0.45 |
| $40^+$–45 | 72 | 0.71 | 0.239 | 0.152 | 71.7 | 0.00 |
| More than 45 | 40 | 1.36 | 0.087 | 0.087 | 41.1 | 0.03 |
| | | | | | Total: | 1.90 |

The 5% significant value for $\chi^2$ with 3 degrees of freedom is 7.8147 and so our value is well within the acceptable limits of statistical error. We conclude that **the normal distribution identified in the solution to Question 1 provides a satisfactory model of the behaviour of the concrete cubes.**

## Life testing

Many of the variables which are monitored in quality control are normally distributed. The lifetime of a component or finished product, however, is often more accurately described by the **Weibull distribution**. This can be written as

$$P(\text{lifetime} \geqslant t) = e^{-\lambda t^m} \tag{10.3}$$

where $m$ and $\lambda$ are parameters appropriate to the particular item whose lifetime is being described. The probability that an item lives longer than time $t$ is the **reliability function** of $t$ or more simply the **reliability**. Engineers find it convenient to measure the **failure rate** of a piece of equipment or an individual part. This is the fraction of all similar systems of the same age which fail at that age for the first time. It can be measured for all ages and has a graph which is typically like Fig. 10.2. The central flat portion defines the useful lifetime of the system and the whole graph is referred to as a **bathtub curve** because of its shape. The failure rate of the Weibull distribution (10.3)

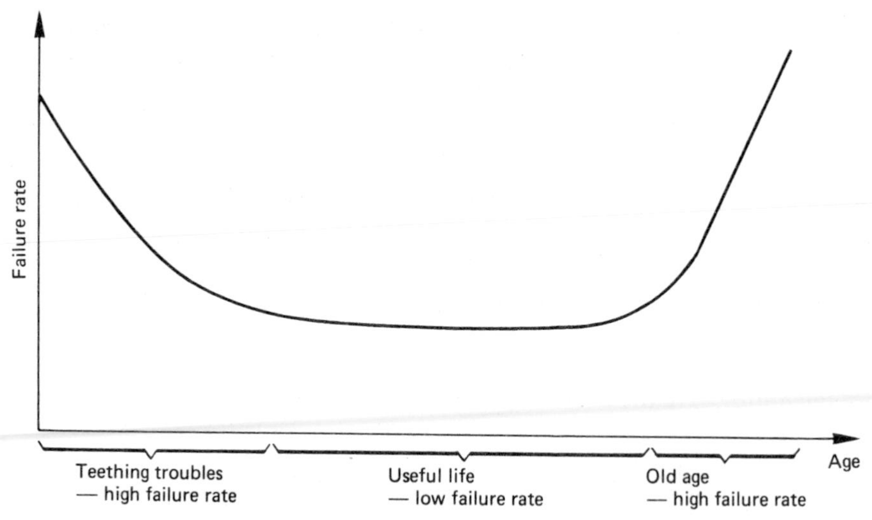


**Fig. 10.2** Typical failure rate against age curve

is $m\lambda t^{m-1}$ at age $t$. The average or expected lifetime is often quoted as well as the failure rate. It is called the **Mean Time Between Failures, MTBF**, or the **Mean Time To Failure, MTTF**.

In the special case where $m$ is equal to 1 the failure rate is equal to $\lambda$ items per unit time period irrespective of the items' age. The failure-rate graph is a horizontal line which is a reasonably good approximation for many systems over their useful life phase. Equation (10.3) reduces to

$$P(\text{lifetime} \geqslant t) = e^{-\lambda t} \qquad (10.4)$$

and is called the **negative exponential distribution**. The expected lifetime is $(1/\lambda)$ for this distribution. Hence the failure rate can be estimated from a sample of lifetimes as 1 divided by the sample mean. In the next example we fit the more general Weibull distribution to the lifetimes of some light bulbs.

### The light bulbs example

A sample of 838 light bulbs were burned for 3000 hours. The numbers still surviving were recorded every 300 hours:

| Time (hours) | 0 | 300 | 600 | 900 | 1200 | 1500 | 1800 | 2100 | 2400 | 2700 | 3000 |
|---|---|---|---|---|---|---|---|---|---|---|---|
| Number surviving | 838 | 611 | 419 | 291 | 194 | 124 | 79 | 47 | 31 | 17 | 9 |

This is called a **life table** or **survival table** for the bulbs. It has the same information content as a grouped frequency table. For instance the number of bulbs whose lifetimes were between 900 and 1200 hours was 291 minus 194 which is 97.

We estimate the most appropriate values of $m$ and $\lambda$ in the Weibull distribution to describe this data.

*Solution* The method is similar to that of the last example. Empirical probabilities are calculated from the data and used to draw a scatter diagram. If the assumed distribution does apply to the variable in question, then the points on this graph will lie in a straight line. The goodness of fit can therefore be judged by the appearance of the diagram and any parameters in the theoretical model estimated from it.

Taking logarithms twice of both sides of equation (10.3), we find that

$$\ln(-\ln P) = \ln \lambda + m(\ln t) \tag{10.5}$$

where 'ln' stands for 'natural logarithm' and is easily obtainable on a scientific calculator. The symbol $P$ represents the probability that the lifetime of an item is longer than or equal to $t$. This equation implies that a graph of the left-hand side against $\ln t$ values should be a straight line. The same reasoning was followed in the Heater Example of Chapter 4.

**Table 10.3** Transformations for the Light Bulbs Example

| Life, $t$ (hours) | Number of bulbs lasting longer | Probability of a bulb lasting longer, $P$ | $\ln(-\ln P)$ | $\ln t$ |
|---|---|---|---|---|
| 0 | 838 | 1.000 | — | — |
| 300 | 611 | 0.729 | −1.15 | 5.70 |
| 600 | 419 | 0.500 | −0.37 | 6.40 |
| 900 | 291 | 0.347 | 0.06 | 6.80 |
| 1200 | 194 | 0.232 | 0.38 | 7.09 |
| 1500 | 124 | 0.148 | 0.65 | 7.31 |
| 1800 | 79 | 0.094 | 0.86 | 7.50 |
| 2100 | 47 | 0.056 | 1.06 | 7.65 |
| 2400 | 31 | 0.037 | 1.19 | 7.78 |
| 2700 | 17 | 0.020 | 1.36 | 7.90 |
| 3000 | 9 | 0.011 | 1.51 | 8.01 |
| | | Total | 5.55 | 72.14 |
| | | Average | 0.555 | 7.214 |

Table 10.3 contains the working needed for the transformation of the data. A graph of the last two columns is shown as Fig. 10.3 and it can be seen that the points are in a clear straight line. As in fitting a normal curve, we can either calculate the equation by a regression analysis or draw a line by inspection through the average point and measure its slope. Adopting the latter strategy we find from the triangle shown in the figure that the slope of the resulting line is 1.16. Substituting the average coordinates into equation (10.5), we obtain

$$0.555 = \ln \lambda + 1.16 \times 7.214 \tag{10.6}$$

Hence $\ln \lambda$ is equal to −7.81324. Using the '$e^x$' key on a calculator to take the antilogarithm, the value of $\lambda$ is therefore about 0.0004. It seems that **the Weibull distribution (10.3) describes the data satisfactorily with $m$ equal to 1.16 and $\lambda$ equal to 0.0004.** A $\chi^2$ goodness-of-fit test could be applied as in the Concrete Example if required.

There are various reasons why a reliability engineer might want to treat the light bulbs as if their lifetimes were negative exponentially distributed. Reference (55) or other operational research texts contain explanations of these motives. The negative exponential distribution corresponds to the Weibull with $m$ equal to 1 and so we could

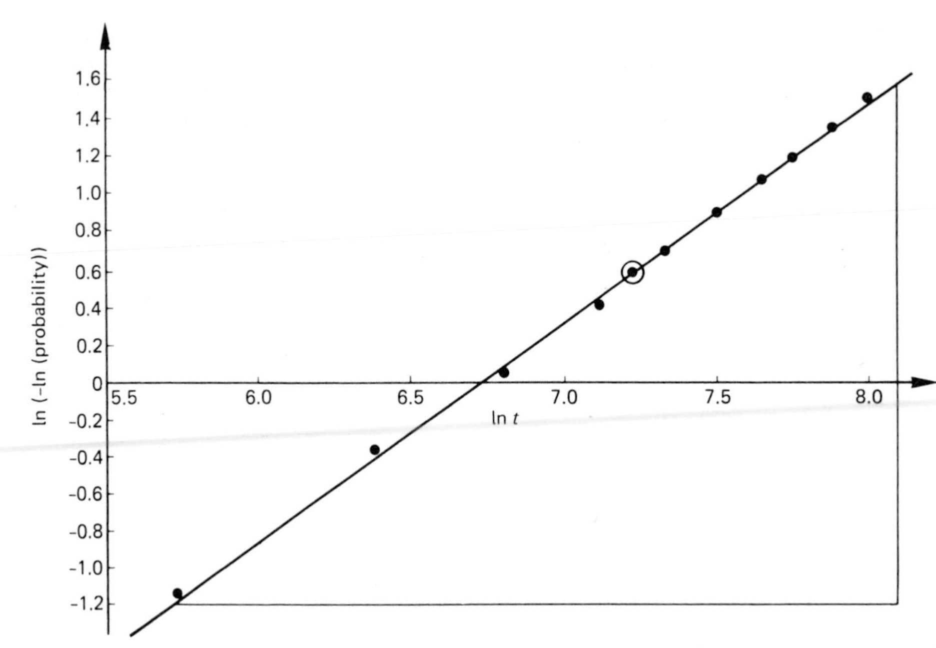


**Fig. 10.3** Regression line for the Light Bulbs Example

test the slope, 1.16, against the null hypothesis value of 1. The procedure would involve calculating the regression line and is covered in the Cholesterol Example of Chapter 4.

## Control charts

A production manager needs to monitor the quality level of the output from a manufacturing process. It can be very costly to produce items, often in bulk, which are unfit for sale or for utilisation elsewhere in the factory. In some processes it is possible to check every single item but usually a small sample of them is taken and examined. Sampling is expensive and may necessitate destroying the items selected. We therefore consider a sequence of samples, typically with just four or five items in each, drawn at fairly regular intervals, possibly daily. When each sample is tested the decision must be made as to whether the process is still '**in control**' or whether it has become '**out of control**'. In the next example we see how this decision-making procedure can be implemented graphically.

### The tablets example

Pharmaceutical tablets are made by combining a measured amount of a drug with an inactive base compound. They are manufactured in large quantities and the weight of drug in each tablet is believed to be normally distributed. On a day when the process is known to be in control 83 tablets are analysed. The average amount of drug per tablet is 280 mg with standard deviation 0.7 mg.

*Question 1* It is proposed to choose 5 tablets from each batch at random and determine the weight of drug in each one. Draw mean and range control charts so that these samples can be used to monitor the quality of the output from the production process.

*Solution* The assumption of normality could be tested as it was for the concrete data earlier in the chapter. Let us suppose that has been done and we are left with good estimates of the population mean and standard deviation, 280 and 0.7 respectively. It is now possible to make some predictions about the behaviour of the samples of 5 tablets which will be drawn on a regular basis.

As argued in the section on Classical Sampling Theory in Chapter 3, the average of a sample drawn from a normal population is itself normally distributed with mean $\mu$ and standard deviation $\sigma/\sqrt{n}$. Applying inequality (3.7) it follows that 95% of samples of 5 tablets will have their average between $280-(1.96\times 0.7/\sqrt{5})$ and $280+(1.96\times 0.7/\sqrt{5})$. This 95% prediction interval for the sample mean, 279.4 mg to 280.6 mg, is based on the population mean being 280 mg. If the sample average does not fall within this interval then we have the usual dilemma of statistical hypothesis testing. Either the observed sample is less than 5% likely to have occurred by chance or the population mean is no longer 280 mg. The extremes of the prediction interval are called **control limits** and we have just calculated the 95% ones. The general formula is

$$\mu \pm \frac{z\sigma}{\sqrt{n}} \qquad (10.7)$$

where $z$ is the appropriate standard normal $z$ score for the percentage required and $n$ is the size of the sample. The plus sign gives the **upper** limit and the minus sign the **lower** limit. Often the $z$ scores 2 and 3 are used to give the **two sigma** and **three sigma** control limits. They correspond to 99.95% and 99.997% prediction intervals respectively and in this case are $280-(2\times 0.7/\sqrt{5})$ to $280+(2\times 0.7/\sqrt{5})$ and $280-(3\times 0.7/\sqrt{5})$ to $280+(3\times 0.7/\sqrt{5})$. These are 279.37 to 280.63 and 279.06 to 280.94.

The nature of the production process being controlled determines which percentage prediction intervals should be used. Similarly it dictates what action is to be taken if a sample average is outside the calculated limits. Usually two sets of limits are incorporated, an **inner** or **warning** pair and an **outer** or **action** pair. These may be 95% and 99% prediction intervals but are more often the less sensitive $2\sigma$ and $3\sigma$ limits. The exact interpretation of the terms 'warning' and 'action' vary but the latter implies some sort of examination of the machinery with a possible total or partial shutdown of the production process.

It is not sufficient to control the average behaviour of the output from a process without monitoring its spread. Contrary to common belief, a statistician is not content with his head in the oven and his feet in the refrigerator because his average temperature is right. A manufacturer of 30 mm bolts is likewise unhappy if the bolt lengths range from 24 mm to 36 mm in giving the average he wants.

The range, which is simply the largest data value minus the smallest, is the most popular measure of spread in quality control as it is the easiest one to calculate. The samples are usually small and so the theoretical objections to range, mainly that it over-responds to outlying sample values, are matched by objections to estimating anything at all from such little data. It can also be argued that the range of the values is a property which is directly relevant to the manufacturer. It represents the difference between extremes in the performance of the production process.

The probability distribution of the range of a sample presents difficulties and percentage points tables are generated by extensive computer calculations. The one shown as Table 10.4 gives upper and lower range control limits as multiples of the population standard deviation. Choosing 95% warning limits and 99% action limits for our samples of 5 tablets we evaluate them to be $0.85\times 0.7$, $4.20\times 0.7$, $0.55\times 0.7$ and $4.89\times 0.7$. These are 0.595, 2.94, 0.385 and 3.423 and we now have two sets of prediction intervals for the sample mean and range.

**Table 10.4** Multipliers of the population standard deviation for range control limits

| Sample size, *n* | Lower control limit multiple | | | | Upper control limit multiple | | | |
|---|---|---|---|---|---|---|---|---|
| | 0.1% | 0.5% | 1.0% | 2.5% | 0.1% | 0.5% | 1.0% | 2.5% |
| 2 | 0.00 | 0.01 | 0.02 | 0.04 | 4.65 | 3.97 | 3.64 | 3.17 |
| 3 | 0.06 | 0.13 | 0.19 | 0.30 | 5.06 | 4.42 | 4.12 | 3.68 |
| 4 | 0.20 | 0.34 | 0.43 | 0.59 | 5.31 | 4.69 | 4.40 | 3.98 |
| 5 | 0.37 | 0.55 | 0.67 | 0.85 | 5.48 | 4.89 | 4.60 | 4.20 |
| 6 | 0.53 | 0.75 | 0.87 | 1.07 | 5.62 | 5.03 | 4.76 | 4.36 |
| 7 | 0.69 | 0.92 | 1.05 | 1.25 | 5.73 | 5.15 | 4.88 | 4.49 |
| 8 | 0.83 | 1.08 | 1.20 | 1.41 | 5.82 | 5.25 | 4.99 | 4.60 |
| 9 | 0.97 | 1.21 | 1.34 | 1.55 | 5.90 | 5.34 | 5.08 | 4.70 |
| 10 | 1.08 | 1.33 | 1.47 | 1.67 | 5.97 | 5.42 | 5.16 | 4.78 |
| 11 | 1.20 | 1.45 | 1.58 | 1.78 | 6.04 | 5.49 | 5.23 | 4.86 |
| 12 | 1.30 | 1.55 | 1.68 | 1.88 | 6.09 | 5.55 | 5.29 | 4.92 |
| More than 12 | See books of tables like reference (66) | | | | | | | |

The implementation of the inspection procedure can be recorded on a chart. The mean and range of each sample is plotted on the chart so that the history of the production process can be seen at a glance. Some decision rules about readjustment of machinery or maintenance relate to these points as a sequence. For instance the '**rule of seven**' states that if seven consecutive points lie on one side of the target mean then the process has effectively gone out of control. This is essentially a sign test as described in Chapter 3 as the probability of a given sample mean being above or below target is 0.5. **Figure 10.4 shows the mean and range charts drawn on the same graph in keeping with common practice.** The data values which have been entered on the chart are from the next question.

*Question 2* The 8 samples chosen subsequent to setting up the control chart gave the following results:

| Sample number | 1 | 2 | 3 | 4 | 5 | 6 | 7 | 8 |
|---|---|---|---|---|---|---|---|---|
| Mean drug weight (mg) | 280.33 | 280.49 | 280.05 | 279.82 | 279.61 | 279.94 | 280.36 | 280.27 |
| Range of weights (mg) | 2.71 | 2.14 | 2.35 | 1.38 | 0.49 | 0.18 | 1.27 | 1.39 |

Plot their behaviour on the control chart and comment.

*Solution* **The averages and ranges are shown on Fig. 10.4. The sample means remain within acceptable limits throughout the 8 time periods but the range becomes unusually small for samples 5 and 6. Provided the mean is in control this does not give cause for concern.**

Students are often bemused that the range control chart should have lower warning and action lines. Surely the smaller the range of values in the sample the better? The

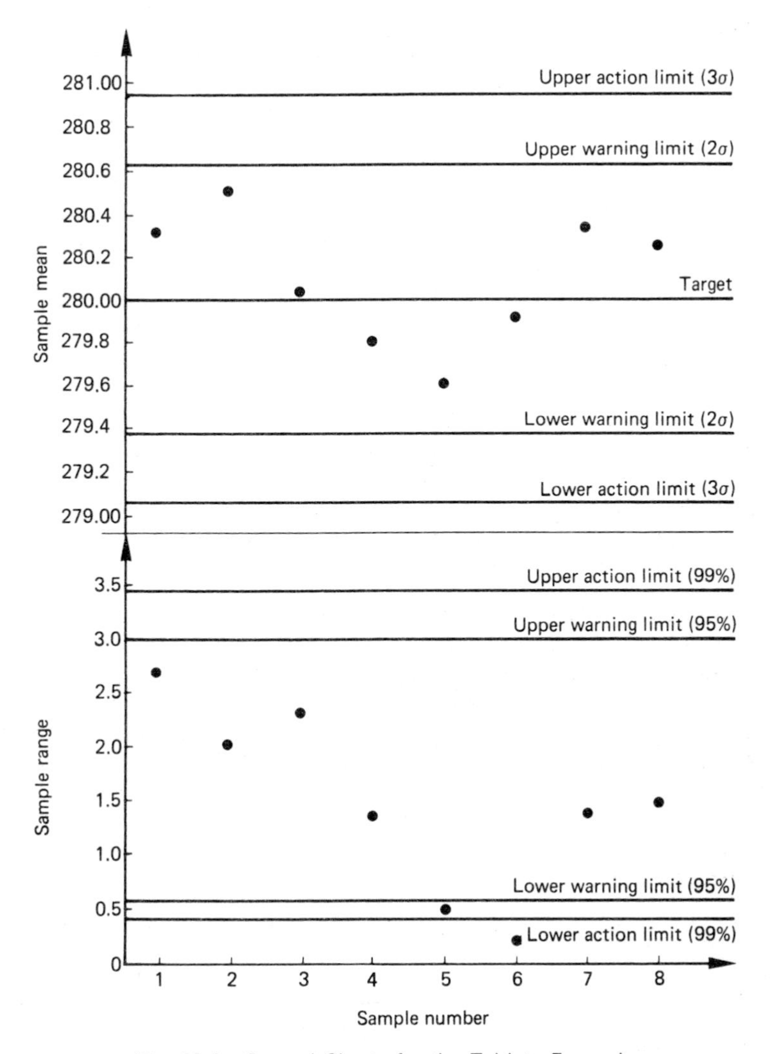


**Fig. 10.4** Control Charts for the Tablets Example

answer to this question is 'yes' only if the mean of the process is still in control. Nobody wants output which is perfectly uniform but at an unacceptable value. The 'action' implied by the lower line of a range chart is simply to inspect the mean chart and ensure that the process is reasonably on target.

We have seen how control charts help to monitor processes for which there is a metric variable whose value acts as an indicator of production quality. In Chapter 3 the distribution of the sample variance was discussed and so a chart for this or the sample standard deviation could be constructed if required. A chart can even be produced for a single value, that is a sample of 1 item. In many manufacturing processes, however, the items being made do not have a metric variable on which quality control can be based. In the electronics industry, for example, items are often 'go' or 'no go' without having any levels of quality between the two. Control charts can be set up for the number of defectives in a sample or the fraction of defectives in a sample using the binomial distribution. The next section deals with the quality control of this type of item not just from the producer's but also from the consumer's point of view.

## Acceptance sampling

Generally an industrial consumer appears on the scene after the items he is buying have been produced. At that time he has no investment in them whereas the manufacturer is relying on their acceptability in order to make a sale. It is not surprising that producer and consumer disagree on whether a particular batch of articles is of good or bad quality. In acceptance sampling a random sample of items is chosen from the batch. The decision to accept the batch is governed by the number of defectives found in the sample.

Whilst we concentrate on items which can be classified as 'good' or 'defective' the methodology is sometimes extended to other kinds of products. For instance a consignment of acid might be tested by examining how many 10 ml specimens out of 5 have the required pH value. Theoretically it is preferable to do parametric tests on metric variables like acidity with control charts or $t$ tests but the acceptance sampling approach is quicker and easier for unskilled personnel to apply.

### The cricket bats example

A department store buys cricket bats in very large batches from a manufacturer abroad. Before accepting a consignment for delivery, quality assurance staff take a random sample of 11 bats and test each one. If there are no more than 2 defectives in the sample then the whole batch is accepted, otherwise it is rejected and sent back to the supplier.

*Question 1* Draw an operating characteristic to show the behaviour of the scheme.

*Solution* An acceptance sampling scheme has two features. There is a **sample size** and a **decision rule**. The sample size here is 11, and the decision rule is to accept the batch if the number of defectives in the sample is less than or equal to the **acceptance number**, 2. In **two-stage** or **multiple sampling** part of the decision rule is 'take another sample' when the number of defectives is in a certain range. **Sequential schemes** involve selecting items for test until a defective is found. The decision is then based on the number of items which had to be chosen before this occurred. In all cases an **operating characteristic** or **o.c.** is a powerful way of analysing the procedure. It is a graph of the probability of accepting a batch against the fraction of defectives which are actually in the batch. The euphemism **quality level** is often used for the fraction of defectives in a batch but we shall refer to it more truthfully as the **fraction defective**.

If a very large consignment of cricket bats has fraction defective $p$, then choosing 11 of them to test corresponds to repeating the same trial 11 times. Every time the probability of 'success', that is finding a defective, is equal to $p$ and the trials are statistically independent of each other. In Chapter 3 this type of situation was described by the binomial distribution. Equation (3.22) with $n$ equal to 11 allows us to calculate the probability of any number of defectives for any fraction defective batch. The probability of accepting the batch can therefore be written as

$$P(\text{accepting batch}) = P(2 \text{ or less defectives}) = P(0) + P(1) + P(2) \qquad (10.8)$$

**Table 10.5** Operating characteristic values for the Cricket Bats Example

| Fraction defective, $p$ | 0.0 | 0.1 | 0.2 | 0.3 | 0.4 | 0.5 | 0.6 | 0.7 | 0.8 | 0.9 | 1.0 |
|---|---|---|---|---|---|---|---|---|---|---|---|
| $P$(accepting batch) | 1.00 | 0.91 | 0.62 | 0.31 | 0.12 | 0.03 | 0.01 | 0.00 | 0.00 | 0.00 | 0.00 |

where the 3 terms on the right are evaluated from equation (3.22) with $r$ put equal to 0, 1 and 2. Hence with the aid of the binomial distribution we are able to determine the probability of accepting a batch with any given value for the fraction defective, $p$.

Table 10.5 shows the result of applying equation (10.8) for values of $p$ ranging from 0 to 1 in steps of 0.1. **Figure 10.5 is the operating characteristic**, which is a graph of the probabilities in this table. It enables the consumer to see how the sampling scheme responds to batches with different levels of the fraction defective. It appears that

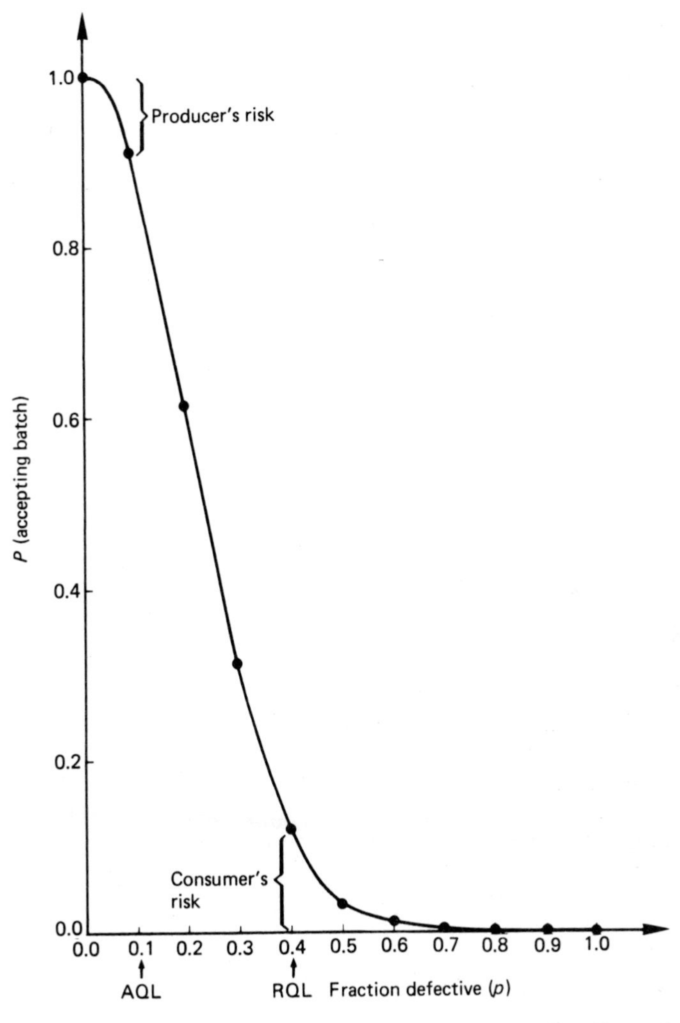


**Fig. 10.5** The operating characteristic for the Cricket Bats Example

batches with fraction defective less than about 0.25 are likely to be accepted while those with a higher level tend to be rejected.

In an ideal world it would not be possible to obtain a misleading sample by chance. Selecting a 'good' sample from a 'bad' batch or conversely a 'bad' sample from a 'good' one would be impossible. Every sample would give perfect information about the batch from which it was drawn. Figure 10.6 is the operating characteristic describing such a Utopian situation. Batches having a fraction defective smaller than the target value are always accepted and those which are worse than the target value are always rejected. An operating characteristic is an important and powerful analytical tool, as besides helping us to identify the types of batches which are being accepted and rejected it shows how close the scheme is to being perfect.

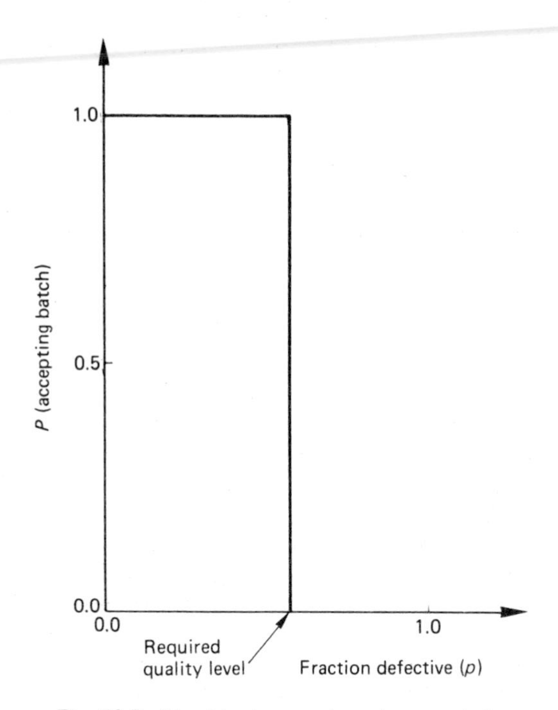


**Fig. 10.6** The ideal operating characteristic

*Question 2* The chief buyer of the department store shudders when told he might be accepting batches in which 40% of the bats are defective. Comment.

*Question* When the consumer specifies a fraction defective in this way it is called the **rejectable quality level, RQL.** The probability of the scheme accepting such a batch is called the **consumer's risk** and from the o.c. we see that it is 0.12. **We can therefore quantify the extent of the chief buyer's shudder and quote the risk as being 12%.**

*Question 3* On one of his rare visits to this country, the manufacturer of the bats expresses horror that the store's quality control is rejecting batches with as few as 10% defectives in them. Comment.

*Solution* The fraction defective specified by the producer as being 'good' is the **acceptable quality level, AQL. We see from the o.c. that this particular scheme imposes a producer's risk on him of 0.09.** It is the probability of such a batch being rejected and is thus 1 minus the probability that it is accepted. In this example if the AQL is 0.1 then his risk is 1 minus 0.91. Notice that whereas the consumer's risk is the probability that the scheme accepts a 'bad' batch, the producer's risk is that it rejects a 'good' one. These correspond to the two types of errors in general statistical hypothesis testing. We run the risk of rejecting a null hypothesis when it is in fact true and of accepting a null hypothesis which is in fact false.

## Summary

**1** Industrial consumers as well as producers are concerned with quality control procedures. In production processes large amounts of data are available, possibly collected in small samples over a period of time. In such conditions a variable can be tested for normality by regression techniques. Similarly a Weibull probability distribution can be fitted to lifetime data. Many components have a constant failure rate during their useful life and a negative exponential distribution, which is a special case of the Weibull, describes their behaviour.

**2** Control charts help a production manager to monitor the average and range of successive samples of items drawn from the manufacturing process. The warning and action lines on them are prediction intervals which may correspond to specific percentages or be $2\sigma$ and $3\sigma$ limits. Control charts can also record the number or fraction of defectives in the samples.

**3** Industrial consumers, and sometimes producers, want to make decisions about batches of items after they have been made, each one being either 'good' or 'defective'. An acceptance sampling scheme has a sample size and a decision rule — which includes the specification of an acceptance number. Batches are classified according to the number of defectives found in the sample. Sequential methods have decision rules based on the number of good items selected before a defective is encountered while multiple-stage sampling schemes have decision rules which require further samples to be taken if the number of defectives is in a borderline region. The performance of acceptance sampling procedures can be analysed by an operating characteristic which is a graph of the probability of accepting a batch against the fraction of defectives in the batch. It can be compared with an ideal curve which has a well-defined acceptable quality level. The producer's risk is the probability that the scheme rejects a batch of acceptable quality while the consumer's risk is the probability that the scheme accepts a batch of rejectable quality.

## Further reading

The fitting of probability distributions to data is covered in many statistics books like references (1) and (3) in the Bibliography. Similarly control charts and acceptance sampling appear often, references (7), (8) and (17) being examples, with (54), (62) and (64) concentrating solely on these topics. The reliability engineering touched on in the chapter is usually considered a branch of operational research because it requires a knowledge of probability theory and is linked with the optimisation of replacement and maintenance strategies. Reference (55) deals with it as do many other operational research texts.

## Exercises

**1** A patient's tolerance to a certain drug is believed to be normally distributed. The numbers of patients with thresholds in various ranges were recorded:

| Threshold (mg per litre) | Less than 14 | $14^{+}$–15 | $15^{+}$–16 | $16^{+}$–17 |
|---|---|---|---|---|
| Number of patients | 20 | 47 | 74 | 92 |

| Threshold (mg per litre) | $17^{+}$–18 | $18^{+}$–19 | $19^{+}$–20 | $20^{+}$–21 | More than 21 |
|---|---|---|---|---|---|
| Number of patients | 110 | 98 | 71 | 60 | 16 |

(i) Fit a normal distribution to the data. (ii) Hence determine the probability that a person chosen at random has a threshold greater than 18.2 mg per litre.
**2** The survival rates of males in Ruritania is:

| Age (years) | 0 | 10 | 20 | 30 | 40 | 50 | 60 | 70 | 80 |
|---|---|---|---|---|---|---|---|---|---|
| Fraction of males surviving | 1.00 | 0.98 | 0.97 | 0.95 | 0.90 | 0.79 | 0.61 | 0.38 | 0.18 |

(i) Fit a Weibull distribution to the data. (ii) Hence calculate the probability that a male Ruritanian chosen at random has a lifetime of between 25 and 47 years.
**3** A machine produces shirt buttons which are supposed to be 12 mm in diameter. A large sample of buttons had a standard deviation of 0.47 mm. Calculate 95% and 99% warning and action limits for mean and range control charts to be used on samples of 6 buttons.
**4** A machine is designed to produce at most 4% defective components. Every day a sample of 8 components is to be tested. Use the binomial distribution to determine how many defectives should be tolerated in the sample before the machine is said to be out of control at the 5% level of significance. Your answer could be used to draw a number defective control chart.
**5** An acceptance sampling scheme consists of choosing a sample of 12 items from a very large batch and accepting the batch if there are 3 or less defectives in the sample. (i) Draw the operating characteristic for the scheme. (ii) Read off from your graph the producer's and consumer's risks if the acceptable quality level is 3% and the rejectable quality level is 28%. (iii) An alternative scheme is proposed with a sample size of 9 and an acceptance number equal to 2. Calculate suitable probabilities to compare this scheme with the original one for both the consumer and producer mentioned in (ii).

# 11 Computing aspects of statistics

Computers are of enormous assistance to statisticians. They give us the ability to attempt a variety of analyses relatively quickly although each one involves complicated calculations on large amounts of data. Any hunch we have about the behaviour of the populations of interest can be followed through within seconds, and the outcome incorporated into our thought processes while the line of reasoning is still fresh. It is the interaction between analyst, computer and data and particularly its immediacy which enables us to acquire a feel for the underlying situation. This facility is especially valuable when analysing multivariate data. Multiple regressions can be carried out at the press of a key with some of the variables mathematically transformed or omitted altogether. As illustrated in Chapter 5, the problem of selecting a suitable set of predictor variables is often resolved by comparing computer analyses of all possible choices with each other.

There are three ways of using computers for statistical processing. Firstly comprehensive **packages** allow us to classify, sort and edit data and perform standard procedures on it. Graphics facilities are usually provided so that scatter diagrams, regression lines, minimal spanning trees and so forth can be seen on the screen. The data is stored as a **worksheet** in which each row corresponds to an item in a sample and each column to a variable measured on that item. This is the format of a data file described in Chapter 9 for the storage of sample survey questionnaire data. Packages are often **command driven** and the computer user types instructions like REGRESS C5 C1 C3 to mean 'perform a regression analysis with column 5 as the dependent variable and columns 1 and 3 as two independent variables'. The result of the calculation and an analysis of variance of its significance would be displayed on the screen, typically within a second.

The second method of applying computer power to a statistical inquiry is to utilise a **spreadsheet package**. A spreadsheet can be thought of as a large piece of paper divided into rows and columns to create **cells**. Each individual cell can be programmed to be the result of a calculation on the contents of other cells. In accountancy applications, some cells act as totals of rows or columns of other cells so that invoices and similar documents can be prepared as the data is entered on the keyboard. Cells may contain words instead of figures so the particular section of the spreadsheet being viewed on the screen contains descriptions to help the operator. These packages are usually **menu driven** and options for drawing graphs and so on are chosen from a menu printed on the screen. Spreadsheets can be designed to implement simple analyses on small samples, like $t$ tests or the analysis of contingency tables. They can be customised by a statistician to satisfy a specific requirement like calculating a confidence interval from a sample drawn daily from a manufacturer's production line. Statistical techniques can also be included in management information systems built with spreadsheets to provide regression forecasts and other operational indicators.

The third way of carrying out statistical calculations by computer is to have purpose-made programs written in a computer language like FORTRAN or Pascal. The time and effort required for this is not usually justified as packages are adequate for most tasks, but in repetitive analyses, like processing questionnaires with the same format every month in a continuing marketing research exercise, the expenditure might be worthwhile. Sometimes it is necessary to convert a data file from one format to another before it can be accessed by a particular package. A 'one off' program could represent a more economical way of overcoming the problem than re-typing all the data.

The usefulness of computers in multivariate analysis as mentioned above warrants a special section in this chapter. A knowledge of Chapter 5 is assumed, together with a familiarity with matrix manipulation. Some multivariate statistics textbooks like Reference (39) in the Bibliography contain an introduction to this topic. The chapter ends by listing and describing programs, written in the computer language BASIC, which evaluate the significance levels and critical values of the various test statistics used throughout this book. Some of the programs enable statistical tables to be derived.

## Multivariate analysis with matrices

*Regression*

We start with an observation matrix **X** in which rows represent cases and columns stand for variables. Thus the element in the 5th row and the 7th column is the value of the 7th variable as measured on the 5th item in the sample. The **transpose** of **X**, written $\mathbf{X}^T$, is formed by taking the first row to be the first column of **X**, the second row to be the second column of **X**, and so on. The matrix $\mathbf{X}^T\mathbf{X}$ is called the **cross-product matrix** as the element in its $i$th row and $j$th column is the sum of products of $x_i$ and $x_j$ across the whole sample. We used these quantities in the Girls' Weights Example of Chapter 5.

Some computer packages insert 1s into the first column of the observation matrix to produce an **augmented observation matrix**, **P**. The first row and column of $\mathbf{P}^T\mathbf{P}$ then have the sum of the data values in them with the size of the sample in the top left-hand corner. If **P** is the augmented observation matrix of just the independent variables $x_1$ and $x_2$ for the Girls' Weights Example, then:

$$\mathbf{P}^T\mathbf{P} = \begin{bmatrix} 8 & 1110 & 74 \\ 1110 & 154652 & 10342 \\ 74 & 10342 & 700 \end{bmatrix} \quad (11.1)$$

This matrix should be compared with the results in Table 5.5 and it can be shown that

$$\mathbf{P}^T\mathbf{P}\begin{bmatrix} a \\ b_1 \\ b_2 \end{bmatrix} = \mathbf{P}^T\mathbf{y} \quad (11.2)$$

Here **y** is the column vector of the 8 observed values for the dependent variable $y$ and $a$, $b_1$ and $b_2$ are as defined by Equations 5.9. Multiplying both sides of this equation by the **inverse matrix** $(\mathbf{P}^T\mathbf{P})^{-1}$ has the effect, by its definition, of making the unknown vector of regression coefficients on the left the subject. A computer program to invert a matrix is given later in this section. We obtain here:

$$\begin{pmatrix} a \\ b_1 \\ b_2 \end{pmatrix} = (\mathbf{P}^T\mathbf{P})^{-1}\mathbf{P}^T\mathbf{y} \quad (11.3)$$

When there are many data values in the sample or they are large, the cross-product matrices $(\mathbf{X}^T\mathbf{X})$ and $(\mathbf{P}^T\mathbf{P})$ have entries which are very big and cause problems. Either the matrix cannot be stored at all because the numbers overflow the memory capacity of the computer or there are unacceptable rounding errors when the matrix is inverted. The situation is eased by subtracting the average of every column of the observation matrix from all the entries in that column. This yields the **centred observation matrix**, which we denote by $\mathbf{Q}$, and $(\mathbf{Q}^T\mathbf{Q})$ is the **corrected cross-products matrix** as its elements are the corrected sums.

The general regression model can be expressed in terms of centred variables:

$$y-\bar{y}=b_1(x_1-\bar{x}_1)+b_2(x_2-\bar{x}_2)+b_3(x_3-\bar{x}_3)+\cdots+b_p(x_p-\bar{x}_p), \tag{11.4}$$

where there are $p$ independent variables. The values of the $b$ parameters are the same as in equations (5.9), which now become:

$$\begin{aligned}
CS_{x_1y} &= b_1CS_{x_1x_1}+b_2CS_{x_1x_2}+b_3CS_{x_1x_3}+\cdots+b_pCS_{x_1x_p}\\
CS_{x_2y} &= b_1CS_{x_2x_1}+b_2CS_{x_2x_2}+b_3CS_{x_2x_3}+\cdots+b_pCS_{x_2x_p}\\
&\vdots\\
CS_{x_py} &= b_1CS_{x_px_1}+b_2CS_{x_px_2}+b_3CS_{x_px_3}+\cdots+b_pCS_{x_px_p}
\end{aligned} \tag{11.5}$$

or in matrix form:

$$\mathbf{Q}^T\mathbf{y}=(\mathbf{Q}^T\mathbf{Q})\mathbf{b} \tag{11.6}$$

The constant $a$ in the model does not appear explicitly in these equations but is simply $(\bar{y}-b_1\bar{x}_1-b_2\bar{x}_2-\cdots-b_p\bar{x}_p)$. The solution of equation (11.6) is

$$\mathbf{b}=(\mathbf{Q}^T\mathbf{Q})^{-1}\mathbf{Q}^T\mathbf{y} \tag{11.7}$$

and the multiple regression coefficient can be expressed in a neat way:

$$R^2=\frac{\mathbf{b}^T\mathbf{Q}^T\mathbf{Q}\mathbf{b}}{\mathbf{y}^T\mathbf{y}} \tag{11.8}$$

In spite of the aesthetic appeal of the centred observation matrix approach, it introduces the possibility of rounding errors right at the beginning of the calculations when the data is centred.

The reader may be puzzled as to why we are preoccupied with accuracy at this point. It is an unfortunate feature of a lot of linear regressions that the independent variables are highly correlated with each other. In the Psychiatry Example of Chapter 5, for instance, the variables measure related medical conditions and in fact the object of the analysis is to put them into similar groups. As a consequence of this almost linear interdependence of the $x$ values, the normal equations (11.5) are often ill-conditioned and the matrices $(\mathbf{X}^T\mathbf{X})$ and $(\mathbf{Q}^T\mathbf{Q})$ difficult to invert accurately. It is advisable to remove variables from the data which are highly correlated with other variables before fitting a regression model. The correlation matrix discussed later helps to identify such variables.

In many applications of multivariate analysis it is desirable to scale all the variables not only to have zero mean, which is achieved by centring, but also to have the same variance. Comparing $z$ scores is preferable to comparing raw data values when the observations are measured in completely diverse systems of units or are of different orders of magnitude. We denote the matrix of observed $z$ scores by $\mathbf{Z}$ so the element in the $i$th row and $j$th column of $\mathbf{Z}$ is the $z$ score of the $i$th sample item on the $j$th variable. Regressing the standardised $y$ values on the $z$ scores is carried out by substituting **Z** for **Q** in equations (11.7) and (11.8). The $k$th regression coefficient is not $b_k$ as in (11.5) but $(s_{x_k} b_k / s_y)$.

For all multivariate work the **correlation matrix** is a useful object to study. The diagonal elements of the corrected cross-products matrix are the corrected sums of the squares of the variables. If $[\text{matrix}]_{ij}$ stands for the entry in the $i$th row and $j$th column of 'matrix' then the correlation matrix **C** is given by:

$$r_{ij} = [\mathbf{C}]_{ij} = \frac{[\mathbf{Q}^{\mathrm{T}}\mathbf{Q}]_{ij}}{\sqrt{([\mathbf{Q}^{\mathrm{T}}\mathbf{Q}]_{ii}[\mathbf{Q}^{\mathrm{T}}\mathbf{Q}]_{jj})}} \qquad (11.9)$$

This follows from equation (4.6) and was the method used to obtain the correlation matrix in the Psychiatry Example. The matrix **C** is also equal to $(\mathbf{Z}^{\mathrm{T}}\mathbf{Z})/(n-1)$.

*Measures of distance*

Distances between cases were defined in Chapter 5 for the Hotels Example and between variables for the Psychiatry Example. Cluster analysis techniques like the drawing of dendrograms and minimal spanning trees require measures of distance to be defined. Now cases can be thought of as points in a space whose coordinate axes are the variables being observed. Similarly a variable can be thought of as a point in a space where the cases are the axes and the coordinates are the values of the variable measured on each case. For instance the case 'Empyre' in the Hotels Example has coordinates (6, 19, 4, no, no) in the space of variables and the variable 'number of bedrooms' has coordinates (37, 20, 150, 90, 19, 48, 138, 142) in the space of cases. The variables are usually scaled into $z$ scores before any distance function between points is considered.

A logical way of specifying what is meant by distance in a multi-dimensional metric space is to extend the concept of straight lines in two-dimensional space. The distance between two points in two dimensions is the hypotenuse of a right-angled triangle whose base is the difference of the points' $x$ coordinates and height is the difference of their $y$ coordinates. It follows from Pythagoras' theorem that the length we want is the square root of the sum of the squared coordinate differences. Generalising this to the multi-dimensional cases and variables spaces gives **Euclidean distance**:

$$\begin{pmatrix}\text{Euclidean distance}\\ \text{between cases } p \text{ and } q\end{pmatrix} = \sqrt{[\text{sum over } k \text{ of } (z_{pk} - z_{qk})^2]}$$
$$= \sqrt{([\mathbf{Z}\mathbf{Z}^{T}]_{pp} + [\mathbf{Z}\mathbf{Z}^{\mathrm{T}}]_{qq} - 2[\mathbf{Z}\mathbf{Z}^{\mathrm{T}}]_{pq})} \qquad (11.10)$$

$$\begin{pmatrix}\text{Euclidean distance}\\ \text{between variables } i \text{ and } j\end{pmatrix} = \sqrt{[\text{sum over } k \text{ of } (z_{ki} - z_{kj})^2]}$$
$$= \sqrt{([\mathbf{Z}^{\mathrm{T}}\mathbf{Z}]_{ii} + [\mathbf{Z}^{\mathrm{T}}\mathbf{Z}]_{jj} - 2[\mathbf{Z}^{\mathrm{T}}\mathbf{Z}]_{ij})}$$
$$= \sqrt{[2(n-1)(1-r_{ij})]} \qquad (11.11)$$

In Chapter 5 we took the distance between variables to be simply 1 minus the correlation coefficient, $r_{ij}$. This gives the same relative positions for the variables as equation (11.11). It is interesting to see the symmetry between cases and variables exhibited by equations (11.10) and (11.11). Replacing **Z** by $\mathbf{Z}^{\mathrm{T}}$ in either one produces the other.

Attribute data can be coded so that the above definitions based on $z$ scores apply. For instance, the presence or absence of certain characteristics in an animal species forms a sequence of attribute variables. Coding 'presence' as '1' and 'absence' as '0' the Euclidean distance (11.10) measures how 'close' two species are to each other. Of course the success of this classification scheme depends on the suitability of the sequence of attributes chosen and the validity of assuming that every one in the sequence has equal weight in the resulting distance.

*Principal components*
An **eigenvector e** of the matrix **M** corresponding to an **eigenvalue** $\lambda$ satisfies the equation:

$$\mathbf{Me} = \lambda \mathbf{e} \tag{11.12}$$

In a principal components analysis we need to find the eigenvalues and eigenvectors of the correlation matrix **C**. The eigenvalues add up to the number of variables and represent the proportion of total variability explained by each component. These ideas were illustrated in the Psychiatry Example. The degree of correlation each variable has with a specific component is given by the vector of correlation coefficients:

$$\text{vector of correlation coefficients} = \frac{\sqrt{\lambda}\,\mathbf{e}}{|\mathbf{e}|} \tag{11.13}$$

The symbol $|\mathbf{e}|$ stands for the length of **e** and is the square root of the sum of the squares of its coordinates. Computer packages usually produce **normalised** eigenvectors which have unit length. The generation of all the eigenvalues and eigenvectors is often performed following a single command from the user.

*Computer programs for matrix manipulation*
It is easy to program the matrix manipulation necessary for statistical analysis. As examples, here are two programs written in the computer language BASIC. One forms the product of the transpose of a matrix with the matrix, that is $\mathbf{M}^T\mathbf{M}$, while the other evaluates the inverse of a matrix.

```
10 PRINT "M TRANSPOSE TIMES M PROGRAM"
20 PRINT "----------------------------"
30 PRINT
40 DIM M(10,10),P(10,10)
50 PRINT "HOW MANY ROWS ";
60 INPUT R
70 PRINT "HOW MANY COLUMNS ";
80 INPUT C
90 FOR I=1 TO R
100 PRINT "ROW ";I;" VALUES, ONE PER LINE"
110 FOR J=1 TO C
120 INPUT M(I,J)
130 NEXT J
140 NEXT I
150 FOR A=1 TO R
160 FOR B=1 TO C
170 P(A,B)=0
180 FOR L=1 TO R
190 P(A,B)=P(A,B)+(M(L,A)*M(L,B))
200 NEXT L
210 PRINT P(A,B);"  ";
220 NEXT B
230 PRINT
240 NEXT A
250 END
```

```
10 PRINT "MATRIX INVERSION PROGRAM"
20 PRINT "------------------------"
30 PRINT
40 DIM M(10,10),V(10,10)
50 PRINT "HOW MANY ROWS ";
60 INPUT R
70 FOR I=1 TO R
80 PRINT "ROW ";I;" VALUES, ONE PER LINE"
90 FOR J= 1 TO R
100 INPUT M(I,J)
110 V(I,J)=0
120 NEXT J
130 V(I,I)=1
140 NEXT I
150 FOR I=1 TO R
160 D=M(I,I)
170 FOR J=1 TO R
180 M(I,J)=M(I,J)/D
190 V(I,J)=V(I,J)/D
200 NEXT J
210 FOR K=1 TO R
220 IF K=I THEN 280
230 F=M(K,I)
240 FOR L=1 TO R
250 M(K,L)=M(K,L)-(F*M(I,L))
260 V(K,L)=V(K,L)-(F*V(I,L))
270 NEXT L
280 NEXT K
290 NEXT I
300 FOR I=1 TO R
310 FOR J=1 TO R
320 PRINT V(I,J);"  ";
330 NEXT J
340 PRINT
350 NEXT I
360 END
```

Both of these programs can process matrices of any size and the DIM instructions should be altered accordingly. They could be merged together as part of a larger program to perform multiple regression.

## Probabilities for test statistics

*The normal distribution*

The area to the right of a $z$ score under the standard normal curve is

$$P(z\text{ score} \geqslant Z) = \int_{Z}^{\infty} \frac{e^{-Z^2/2}}{\sqrt{(2\pi)}} \mathrm{d}Z$$
$$= 0.5 - \frac{1}{\sqrt{(2\pi)}}\left(Z - \frac{Z^3}{3.2} + \frac{Z^5}{5.2^2.2!} - \frac{Z^7}{7.2^3.3!} + \cdots\right) \quad (11.14)$$

**Table 11.1** Area under the Standard Normal Curve

Area tabulated

z

| z | .ØØ | .Ø1 | .Ø2 | .Ø3 | .Ø4 | .Ø5 | .Ø6 | .Ø7 | .Ø8 | .Ø9 |
|---|---|---|---|---|---|---|---|---|---|---|
| Ø.Ø | .5ØØØØ | .496Ø1 | .492Ø2 | .488Ø3 | .484Ø5 | .48ØØ6 | .476Ø8 | .4721Ø | .46812 | .46414 |
| Ø.1 | .46Ø17 | .4562Ø | .45224 | .44828 | .44433 | .44Ø38 | .43644 | .43251 | .42858 | .42465 |
| Ø.2 | .42Ø74 | .41683 | .41294 | .4Ø9Ø5 | .4Ø517 | .4Ø129 | .39743 | .39358 | .38974 | .38591 |
| Ø.3 | .382Ø9 | .37828 | .37448 | .37Ø7Ø | .36693 | .36317 | .35942 | .35569 | .35197 | .34827 |
| Ø.4 | .34458 | .34Ø9Ø | .33724 | .3336Ø | .32997 | .32636 | .32276 | .31918 | .31561 | .312Ø7 |
| Ø.5 | .3Ø854 | .3Ø5Ø3 | .3Ø153 | .298Ø6 | .2946Ø | .29116 | .28774 | .28434 | .28Ø96 | .2776Ø |
| Ø.6 | .27425 | .27Ø93 | .26763 | .26435 | .261Ø9 | .25785 | .25463 | .25143 | .24825 | .2451Ø |
| Ø.7 | .24196 | .23885 | .23576 | .2327Ø | .22965 | .22663 | .22363 | .22Ø65 | .2177Ø | .21476 |
| Ø.8 | .21186 | .2Ø897 | .2Ø611 | .2Ø327 | .2ØØ45 | .19766 | .19489 | .19215 | .18943 | .18673 |
| Ø.9 | .184Ø6 | .18141 | .17879 | .17619 | .17361 | .171Ø6 | .16853 | .166Ø2 | .16354 | .161Ø9 |
| 1.Ø | .15866 | .15625 | .15386 | .1515Ø | .14917 | .14686 | .14457 | .14231 | .14ØØ7 | .13786 |
| 1.1 | .13567 | .1335Ø | .13136 | .12924 | .12714 | .125Ø7 | .123Ø2 | .121ØØ | .119ØØ | .117Ø2 |
| 1.2 | .115Ø7 | .11314 | .11123 | .1Ø935 | .1Ø749 | .1Ø565 | .1Ø383 | .1Ø2Ø4 | .1ØØ27 | .Ø9853 |
| 1.3 | .Ø968Ø | .Ø951Ø | .Ø9342 | .Ø9176 | .Ø9Ø12 | .Ø8851 | .Ø8691 | .Ø8534 | .Ø8379 | .Ø8226 |
| 1.4 | .Ø8Ø76 | .Ø7927 | .Ø778Ø | .Ø7636 | .Ø7493 | .Ø7353 | .Ø7214 | .Ø7Ø78 | .Ø6944 | .Ø6811 |
| 1.5 | .Ø6681 | .Ø6552 | .Ø6426 | .Ø63Ø1 | .Ø6178 | .Ø6Ø57 | .Ø5938 | .Ø5821 | .Ø57Ø5 | .Ø5592 |
| 1.6 | .Ø548Ø | .Ø537Ø | .Ø5262 | .Ø5155 | .Ø5Ø5Ø | .Ø4947 | .Ø4846 | .Ø4746 | .Ø4648 | .Ø4551 |
| 1.7 | .Ø4457 | .Ø4363 | .Ø4272 | .Ø4182 | .Ø4Ø93 | .Ø4ØØ6 | .Ø392Ø | .Ø3836 | .Ø3754 | .Ø3673 |
| 1.8 | .Ø3593 | .Ø3515 | .Ø3438 | .Ø3362 | .Ø3288 | .Ø3216 | .Ø3144 | .Ø3Ø74 | .Ø3ØØ5 | .Ø2938 |
| 1.9 | .Ø2872 | .Ø28Ø7 | .Ø2743 | .Ø268Ø | .Ø2619 | .Ø2559 | .Ø25ØØ | .Ø2442 | .Ø2385 | .Ø233Ø |
| 2.Ø | .Ø2275 | .Ø2222 | .Ø2169 | .Ø2118 | .Ø2Ø68 | .Ø2Ø18 | .Ø197Ø | .Ø1923 | .Ø1876 | .Ø1831 |
| 2.1 | .Ø1786 | .Ø1743 | .Ø17ØØ | .Ø1659 | .Ø1618 | .Ø1578 | .Ø1539 | .Ø15ØØ | .Ø1463 | .Ø1426 |
| 2.2 | .Ø139Ø | .Ø1355 | .Ø1321 | .Ø1287 | .Ø1255 | .Ø1222 | .Ø1191 | .Ø116Ø | .Ø113Ø | .Ø11Ø1 |
| 2.3 | .Ø1Ø72 | .Ø1Ø44 | .Ø1Ø17 | .ØØ99Ø | .ØØ964 | .ØØ939 | .ØØ914 | .ØØ889 | .ØØ866 | .ØØ842 |
| 2.4 | .ØØ82Ø | .ØØ798 | .ØØ776 | .ØØ755 | .ØØ734 | .ØØ714 | .ØØ695 | .ØØ676 | .ØØ657 | .ØØ639 |
| 2.5 | .ØØ621 | .ØØ6Ø4 | .ØØ587 | .ØØ57Ø | .ØØ554 | .ØØ539 | .ØØ523 | .ØØ5Ø9 | .ØØ494 | .ØØ48Ø |
| 2.6 | .ØØ466 | .ØØ453 | .ØØ44Ø | .ØØ427 | .ØØ415 | .ØØ4Ø3 | .ØØ391 | .ØØ379 | .ØØ368 | .ØØ357 |
| 2.7 | .ØØ347 | .ØØ336 | .ØØ326 | .ØØ317 | .ØØ3Ø7 | .ØØ298 | .ØØ289 | .ØØ28Ø | .ØØ272 | .ØØ264 |
| 2.8 | .ØØ256 | .ØØ248 | .ØØ24Ø | .ØØ233 | .ØØ226 | .ØØ219 | .ØØ212 | .ØØ2Ø5 | .ØØ199 | .ØØ193 |
| 2.9 | .ØØ187 | .ØØ181 | .ØØ175 | .ØØ169 | .ØØ164 | .ØØ159 | .ØØ154 | .ØØ149 | .ØØ144 | .ØØ139 |
| 3.Ø | .ØØ135 | .ØØ131 | .ØØ126 | .ØØ122 | .ØØ118 | .ØØ114 | .ØØ111 | .ØØ1Ø7 | .ØØ1Ø4 | .ØØ1ØØ |
| 3.1 | .ØØØ97 | .ØØØ94 | .ØØØ9Ø | .ØØØ87 | .ØØØ85 | .ØØØ82 | .ØØØ79 | .ØØØ76 | .ØØØ74 | .ØØØ71 |
| 3.2 | .ØØØ69 | .ØØØ66 | .ØØØ64 | .ØØØ62 | .ØØØ6Ø | .ØØØ58 | .ØØØ56 | .ØØØ54 | .ØØØ52 | .ØØØ5Ø |
| 3.3 | .ØØØ48 | .ØØØ47 | .ØØØ45 | .ØØØ43 | .ØØØ42 | .ØØØ4Ø | .ØØØ39 | .ØØØ38 | .ØØØ36 | .ØØØ35 |
| 3.4 | .ØØØ34 | .ØØØ32 | .ØØØ31 | .ØØØ3Ø | .ØØØ29 | .ØØØ28 | .ØØØ27 | .ØØØ26 | .ØØØ25 | .ØØØ24 |
| 3.5 | .ØØØ23 | .ØØØ22 | .ØØØ22 | .ØØØ21 | .ØØØ2Ø | .ØØØ19 | .ØØØ19 | .ØØØ18 | .ØØØ17 | .ØØØ17 |
| 3.6 | .ØØØ16 | .ØØØ15 | .ØØØ15 | .ØØØ14 | .ØØØ14 | .ØØØ13 | .ØØØ13 | .ØØØ12 | .ØØØ12 | .ØØØ11 |
| 3.7 | .ØØØ11 | .ØØØ1Ø | .ØØØ1Ø | .ØØØ1Ø | .ØØØØ9 | .ØØØØ9 | .ØØØØ9 | .ØØØØ8 | .ØØØØ8 | .ØØØØ8 |
| 3.8 | .ØØØØ7 | .ØØØØ7 | .ØØØØ7 | .ØØØØ6 | .ØØØØ6 | .ØØØØ6 | .ØØØØ6 | .ØØØØ5 | .ØØØØ5 | .ØØØØ5 |
| 3.9 | .ØØØØ5 | .ØØØØ5 | .ØØØØ4 | .ØØØØ4 | .ØØØØ4 | .ØØØØ4 | .ØØØØ4 | .ØØØØ4 | .ØØØØ3 | .ØØØØ3 |

**Table 11.2** Percentage points of the standard normal distribution

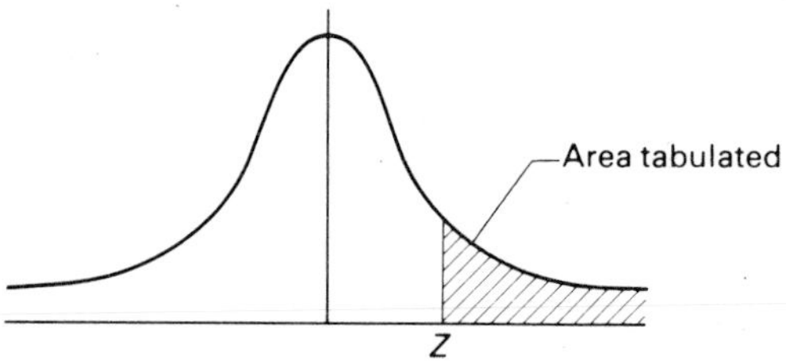

| AREA TO RIGHT | Z SCORE |
|---|---|
| 0.00 | 4.86168 |
| 0.005 | 2.57583 |
| 0.01 | 2.32636 |
| 0.02 | 2.05376 |
| 0.025 | 1.96000 |
| 0.03 | 1.88079 |
| 0.04 | 1.75069 |
| 0.05 | 1.64485 |
| 0.06 | 1.55477 |
| 0.07 | 1.47579 |
| 0.08 | 1.40507 |
| 0.09 | 1.34076 |
| 0.10 | 1.28155 |
| 0.11 | 1.22653 |
| 0.12 | 1.17499 |
| 0.13 | 1.12639 |
| 0.14 | 1.08032 |
| 0.15 | 1.03643 |
| 0.16 | 0.99446 |
| 0.17 | 0.95417 |
| 0.18 | 0.91537 |
| 0.19 | 0.87790 |
| 0.20 | 0.84162 |
| 0.21 | 0.80642 |
| 0.22 | 0.77219 |
| 0.23 | 0.73885 |
| 0.24 | 0.70630 |
| 0.25 | 0.67449 |
| 0.26 | 0.64335 |
| 0.27 | 0.61281 |
| 0.28 | 0.58284 |
| 0.29 | 0.55338 |
| 0.30 | 0.52440 |
| 0.31 | 0.49585 |
| 0.32 | 0.46770 |
| 0.33 | 0.43991 |
| 0.34 | 0.41246 |
| 0.35 | 0.38532 |
| 0.36 | 0.35846 |
| 0.37 | 0.33185 |
| 0.38 | 0.30548 |
| 0.39 | 0.27932 |
| 0.40 | 0.25335 |
| 0.41 | 0.22755 |
| 0.42 | 0.20189 |
| 0.43 | 0.17637 |
| 0.44 | 0.15097 |
| 0.45 | 0.12566 |
| 0.46 | 0.10043 |
| 0.47 | 0.07527 |
| 0.48 | 0.05015 |
| 0.49 | 0.02507 |

The power series is obtained by integrating the Maclaurin series for the integrand between 0 and $Z$. This is subtracted from the total area from zero to infinity which is 0.5. This program sums the series until the terms are less than or equal to 0.00001:

```
10 PRINT "STANDARD NORMAL PROBABILITIES"
20 PRINT "-----------------------------"
30 PRINT
40 PRINT "Z SCORE ";
50 INPUT Z
60 T=Z
70 S=Z
80 R=1
90 T=(-T*Z*Z*(2*R-1))
100 T=T/((2*R+1)*2*R)
110 S=S+T
120 R=R+1
130 IF ABS(T)>.00001 THEN 90
140 A=.5-(S/SQR(8*ATN(1)))
150 PRINT "PROBABILITY OF A Z SCORE GREATER THAN ";Z;" IS ";A
160 END
```

The program can be used to test the significance of a $z$ score, for instance entering the value $-2.1$ for $Z$ should produce the probability 0.98214. When included in a succession of FOR loops to organise the necessary repetition, Table 11.1 can be generated.

The next program runs the last one on a trial-and-error basis to produce a $Z$ value which results in a given area to the right, what we have called a percentage point. The numerical procedure is known as the binary chop and consists of halving a range of values within which the answer is known to be. The new range is taken to be the half containing the answer and the process is repeated. Thus a sequence of ranges is calculated, each half the size of its predecessor and all of them containing the answer. When the area resulting from the mid-point of one of these ranges is sufficiently close to the given area then the algorithm stops and the mid-point is printed out.

```
10 PRINT "NORMAL DISTRIBUTION PERCENTAGE POINTS"
20 PRINT "-------------------------------------"
30 PRINT
40 PRINT "AREA TO THE RIGHT ";
50 INPUT P
60 X=-5
70 Y=5
80 Z=(X+Y)/2
90 T=Z
100 S=Z
110 R=1
120 T=(-T*Z*Z*(2*R-1))
130 T=T/((2*R+1)*2*R)
140 S=S+T
150 R=R+1
160 IF ABS(T)>.00001 THEN 120
170 A=.5-(S/SQR(8*ATN(1)))
```

```
180 IF A<P THEN Y=Z
190 IF A>=P THEN X=Z
200 IF (Y-X)>.000001 THEN 80
210 PRINT "Z SCORE IS "; Z
220 END
```

Like the previous one, this program can also be used for 'one-off' values of significant $z$ scores. However, putting it inside a FOR loop enables a table of critical $z$ scores to be printed as in Table 11.2.

*The chi-squared distribution*

The area under its distribution curve to the right of a chi-squared value $X$ with $V$ degrees of freedom is

$$P(\chi^2 \geq X) = \frac{\int_X^\infty X^{(V-2)/2}\,e^{-X/2}\,dX}{\int_0^\infty X^{(V-2)/2}\,e^{-X/2}\,dX}$$

$$= 2\,(\text{area to right of } \sqrt{X} \text{ on standard normal curve})$$

$$+\frac{e^{-X/2}}{\sqrt{\pi}}\left(\frac{(X/2)^{1/2}}{\frac{1}{2}}+\frac{(X/2)^{3/2}}{\frac{3}{2}\cdot\frac{1}{2}}+\frac{(X/2)^{5/2}}{\frac{5}{2}\cdot\frac{3}{2}\cdot\frac{1}{2}}+\cdots+\frac{(X/2)^{(V/2)-1}}{(V/2-1)(V/2-2)\cdots\frac{1}{2}}\right)$$

if $V$ is odd

$$= e^{-X/2}\left(1+\frac{(X/2)^1}{1!}+\frac{(X/2)^2}{2!}+\cdots+\frac{(X/2)^{(V/2)-1}}{(V/2-1)!}\right) \qquad (11.15)$$

if $V$ is even.

The derivation of the power series expansions involves integration by parts and the use of reduction formulae. They are much easier to program a computer with than they appear to be, the only complication being to test whether $V$ is even or odd. Here is a program which implements them:

```
10 PRINT "CHI-SQUARED DISTRIBUTION PROBABILITIES"
20 PRINT "-----------------------------------"
30 PRINT
40 PRINT "VALUE OF CHI-SQUARED ";
50 INPUT X
60 IF X<0 THEN 40
70 PRINT "HOW MANY DEGREES OF FREEDOM ";
80 INPUT V
90 IF V>100 THEN 70
100 IF V<0 THEN 70
110 IF V<>INT(V) THEN 70
120 H=(V/2)-INT(V/2)
130 A=EXP(-X/2)
140 T=A
150 R=1
160 IF H=.5 THEN GOSUB 240
170 IF R>(V/2)-1 THEN 220
180 T=T*X/(2*R)
190 A=A+T
200 R=R+1
210 GOTO 170
```

```
220 PRINT "PROBABILITY OF CHI-SQUARED GREATER THAN ";X;" IS ";A
230 END
240 A=0
250 IF X>25 THEN 340
260 T1=SQR(X)
270 R=1
280 A=T1
290 T1=-T1*X/(2*R)
300 A=A+(T1/(2*R+1))
310 R=R+1
320 IF ABS(T1)>.000001 THEN 290
330 A=1-(A/SQR(2*ATN(1)))
340 T=EXP(-X/2)/SQR(2*X*ATN(1))
350 R=.5
360 RETURN
```

As for the normal distribution program, this one can be made the subject of a binary chop procedure to search for percentage points:

```
10 PRINT "CHI-SQUARED DISTRIBUTION PERCENTAGE POINTS"
20 PRINT "------------------------------------------"
30 PRINT
40 PRINT "AREA IN RIGHT HAND TAIL AS A DECIMAL ";
50 INPUT P
60 IF P<.001 THEN 40
70 IF P>1 THEN 40
80 PRINT "HOW MANY DEGREES OF FREEDOM ";
90 INPUT V
100 IF V>100 THEN 80
110 IF V<0 THEN 80
120 IF V<>INT(V) THEN 80
130 H=(V/2)-INT(V/2)
140 X=0
150 Y=150
160 Z=(X+Y)/2
170 A=EXP(-Z/2)
180 T=A
190 R=1
200 IF H=.5 THEN GOSUB 310
210 IF R>(V/2)-1 THEN 260
220 T=T*Z/(2*R)
230 A=A+T
240 R=R+1
250 GOTO 210
260 IF A<P THEN Y=Z
270 IF A>=P THEN X=Z
280 IF (Y-X)>.00005 THEN 160
290 PRINT "CRITICAL CHI-SQUARED VALUE IS ";Z
300 END
310 A=0
320 IF Z>25 THEN 410
```

**Table 11.3** Percentage points of the chi-squared distribution

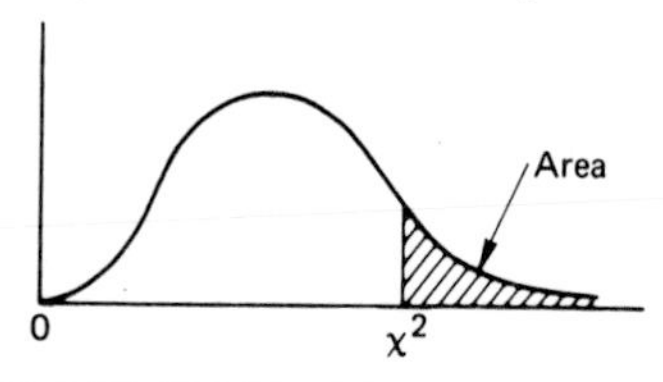


| Degrees of freedom, $\nu$ | Area to the right | | | |
|---|---|---|---|---|
| | 0.975 | 0.050 | 0.025 | 0.010 |
| 1 | 0.0000 | 3.8415 | 5.0239 | 6.6349 |
| 2 | 0.0506 | 5.9915 | 7.3778 | 9.2103 |
| 3 | 0.2158 | 7.8147 | 9.3484 | 11.3449 |
| 4 | 0.4844 | 9.4877 | 11.1433 | 13.2767 |
| 5 | 0.8312 | 11.0705 | 12.8325 | 15.0863 |
| 6 | 1.2373 | 12.5916 | 14.4494 | 16.8119 |
| 7 | 1.6899 | 14.0671 | 16.0128 | 18.4753 |
| 8 | 2.1797 | 15.5073 | 17.5345 | 20.0902 |
| 9 | 2.7004 | 16.9190 | 19.0228 | 21.6660 |
| 10 | 3.2470 | 18.3070 | 20.4832 | 23.2093 |
| 11 | 3.8158 | 19.6751 | 21.9200 | 24.7250 |
| 12 | 4.4038 | 21.0261 | 23.3367 | 26.2170 |
| 13 | 5.0088 | 22.3620 | 24.7356 | 27.6882 |
| 14 | 5.6287 | 23.6848 | 26.1189 | 29.1412 |
| 15 | 6.2621 | 24.9958 | 27.4884 | 30.5779 |
| 16 | 6.9077 | 26.2962 | 28.8454 | 31.9999 |
| 17 | 7.5642 | 27.5871 | 30.1910 | 33.4087 |
| 18 | 8.2308 | 28.8693 | 31.5264 | 34.8053 |
| 19 | 8.9065 | 30.1435 | 32.8523 | 36.1909 |
| 20 | 9.5908 | 31.4104 | 34.1696 | 37.5662 |
| 21 | 10.2829 | 32.6706 | 35.4789 | 38.9322 |
| 22 | 10.9823 | 33.9244 | 36.7807 | 40.2894 |
| 23 | 11.6886 | 35.1725 | 38.0756 | 41.6384 |
| 24 | 12.4012 | 36.4150 | 39.3641 | 42.9798 |
| 25 | 13.1197 | 37.6525 | 40.6465 | 44.3141 |
| 26 | 13.8439 | 38.8851 | 41.9232 | 45.6417 |
| 27 | 14.5734 | 40.1133 | 43.1945 | 46.9629 |
| 28 | 15.3079 | 41.3371 | 44.4608 | 48.2782 |
| 29 | 16.0471 | 42.5570 | 45.7223 | 49.5879 |
| 30 | 16.7908 | 43.7730 | 46.9792 | 50.8922 |

For values of $\nu$ bigger than 30, the critical $\chi^2$ is approximately $\frac{1}{2}(z+\sqrt{(2\nu-1)})^2$ where $z$ is the appropriate one-tail $z$ score.

*Reminder*: Yates' correction should be used if $\nu=1$.

```
330 T1=SQR(Z)
340 R=1
350 A=T1
360 T1=-T1*Z/(2*R)
370 A=A+(T1/(2*R+1))
380 R=R+1
390 IF ABS(T1)>.000001 THEN 360
400 A=1-(A/SQR(2*ATN(1)))
410 T=EXP(-Z/2)/SQR(2*Z*ATN(1))
420 R=.5
430 RETURN
```

Table 11.3 is the result of repeating the execution of this program for all different percentage probabilities. It is obviously advisable to arrange the repetition as a FOR loop or equivalent.

*The F distribution*
The area under its distribution curve to the right of the ratio $F$ of a variance with $V_1$ degrees of freedom divided by a variance with $V_2$ degrees of freedom is

$$P(F \geq X) = \frac{\int_0^W W^{(V_2/2)-1}(1-W)^{(V_1/2)-1}\,\mathrm{d}W}{\int_0^1 W^{(V_2/2)-1}(1-W)^{(V_1/2)-1}\,\mathrm{d}W} \tag{11.16}$$

where $W = V_2/(V_2 + V_1 X)$.

The integral in the numerator can be expressed as a series of powers of $W$:

$$\frac{W^{(V_2/2)-1}}{V_1/2}(1-(1-W)^{V_1/2}) - \frac{((V_2/2)-1)\,W^{V_2/2}}{V_2/2} + \frac{((V_2/2)-1)((V_1/2)-1)\,W^{(V_2/2)+1}}{((V_2/2)+1)2!}$$
$$- \frac{((V_2/2)-1)((V_1/2)-1)((V_1/2)-2)\,W^{V_2/2}}{((V_2/2)+2)3!} + \cdots \tag{11.17}$$

In the program below the summation of (11.17) is written as a subroutine. It is called with $W=1$ to give the denominator of (11.16) and then with the value necessary to calculate the numerator.

```
10 PRINT "F DISTRIBUTION PROBABILITIES"
20 PRINT "----------------------------"
30 PRINT
40 PRINT "F VALUE ";
50 INPUT X
60 PRINT "DEGREES OF FREEDOM OF NUMERATOR ";
70 GOSUB 310
80 V1=V
90 U1=V1/2
100 PRINT "DEGREES OF FREEDOM OF DENOMINATOR ";
110 GOSUB 310
120 V2=V
130 U2=V2/2
140 W=1
150 GOSUB 220
160 B=S
170 W=V2/(V2+(V1*X))
```

```
180 GOSUB 220
190 A=S/B
200 PRINT "PROBABILITY OF AN F RATIO GREATER THAN ";X;" IS ";A
210 END
220 S=((W^(U2-1))*(1-(1-W)^U1)/U1)-(((U2-1)*(W^U2))/U2)
230 D=-(W^U2)*(U2-1)
240 C=1
250 D=-D*(U1-C)*W/(C+1)
260 T=D/(U2+C)
270 S=S+T
280 C=C+1
290 IF ABS(T)>1E-08 THEN 250
300 RETURN
310 INPUT V
320 IF V<1 THEN 350
330 IF INT(V)<>V THEN 350
340 RETURN
350 PRINT "INVALID DATA"
360 STOP
```

As for the last two distributions, we can embed this program into a search algorithm to find percentage points. The program and corresponding 5% significance tables are given below. For larger numbers of degrees of freedom the program needs to run to a high level of numerical accuracy. Many computers allow double precision arithmetic for this purpose. The limit in line 370 on the size of the terms in the series (11.17) can also be reduced so that more of them are included in the summation.

```
10 PRINT "F DISTRIBUTION PERCENTAGE POINTS"
20 PRINT "--------------------------------"
30 PRINT
40 PRINT "AREA IN RIGHT HAND TAIL AS A DECIMAL ";
50 INPUT P
60 IF P<.01 THEN 30
70 IF P>1 THEN 30
80 PRINT "DEGREES OF FREEDOM OF NUMERATOR ";
90 GOSUB 390
100 V1=V
110 U1=V1/2
120 PRINT "DEGREES OF FREEDOM OF DENOMINATOR ";
130 GOSUB 390
140 V2=V
150 U2=V2/2
160 W=1
170 GOSUB 300
180 B=S
190 L=0
200 R=100
210 X=(L+R)/2
220 W=V2/(V2+(V1*X))
230 GOSUB 300
240 A=S/B
250 IF A<P THEN R=X
```

```
260 IF A>=P THEN L=X
270 IF (R-L)>.0004 THEN 210
280 PRINT "CRITICAL F VALUE IS "; X
290 END
300 S=((W^(U2-1))*(1-(1-W)^U1)/U1)-(((U2-1)*(W^U2))/U2)
310 D=-(W^U2)*(U2-1)
320 C=1
330 D=-D*(U1-C)*W/(C+1)
340 T=D/(U2+C)
350 S=S+T
360 C=C+1
```

**Table 11.4** 5% percentage points for the *F* distribution

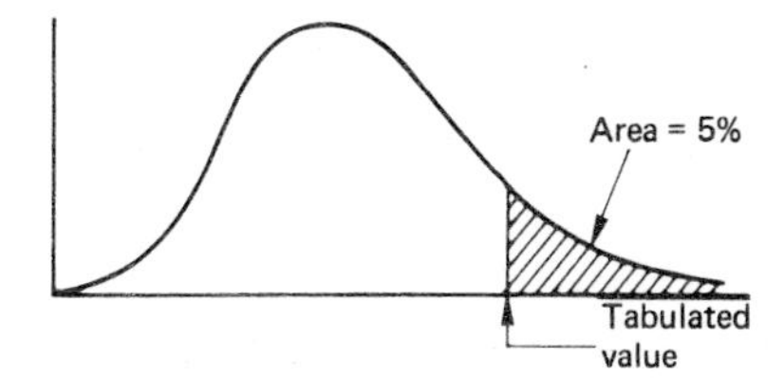


| $\nu_2$ \ $\nu_1$ | 1 | 2 | 3 | 4 | 5 | 6 | 7 | 8 | 9 | 10 | 15 | 20 | 40 |
|---|---|---|---|---|---|---|---|---|---|---|---|---|---|
| 2 | 18.51 | 19.00 | 19.16 | 19.25 | 19.30 | 19.33 | 19.35 | 19.37 | 19.38 | 19.40 | 19.43 | 19.45 | 19.47 |
| 3 | 10.13 | 9.55 | 9.28 | 9.12 | 9.01 | 8.94 | 8.89 | 8.85 | 8.81 | 8.79 | 8.70 | 8.66 | 8.59 |
| 4 | 7.71 | 6.94 | 6.59 | 6.39 | 6.26 | 6.16 | 6.09 | 6.04 | 6.00 | 5.96 | 5.86 | 5.80 | 5.72 |
| 5 | 6.61 | 5.79 | 5.41 | 5.19 | 5.05 | 4.95 | 4.88 | 4.82 | 4.77 | 4.74 | 4.62 | 4.56 | 4.46 |
| 6 | 5.99 | 5.14 | 4.76 | 4.53 | 4.39 | 4.28 | 4.21 | 4.15 | 4.10 | 4.06 | 3.94 | 3.87 | 3.77 |
| 7 | 5.59 | 4.74 | 4.35 | 4.12 | 3.97 | 3.87 | 3.79 | 3.73 | 3.68 | 3.64 | 3.51 | 3.44 | 3.34 |
| 8 | 5.32 | 4.46 | 4.07 | 3.84 | 3.69 | 3.58 | 3.50 | 3.44 | 3.39 | 3.35 | 3.22 | 3.15 | 3.04 |
| 9 | 5.12 | 4.26 | 3.86 | 3.63 | 3.48 | 3.37 | 3.29 | 3.23 | 3.18 | 3.14 | 3.01 | 2.94 | 2.83 |
| 10 | 4.96 | 4.10 | 3.71 | 3.48 | 3.33 | 3.22 | 3.14 | 3.07 | 3.02 | 2.98 | 2.85 | 2.77 | 2.66 |
| 11 | 4.84 | 3.98 | 3.59 | 3.36 | 3.20 | 3.09 | 3.01 | 2.95 | 2.90 | 2.85 | 2.72 | 2.65 | 2.53 |
| 12 | 4.75 | 3.89 | 3.49 | 3.26 | 3.11 | 3.00 | 2.91 | 2.85 | 2.80 | 2.75 | 2.62 | 2.54 | 2.43 |
| 13 | 4.67 | 3.81 | 3.41 | 3.18 | 3.03 | 2.92 | 2.83 | 2.77 | 2.71 | 2.67 | 2.53 | 2.46 | 2.34 |
| 14 | 4.60 | 3.74 | 3.34 | 3.11 | 2.96 | 2.85 | 2.76 | 2.70 | 2.65 | 2.60 | 2.46 | 2.39 | 2.27 |
| 15 | 4.54 | 3.68 | 3.29 | 3.06 | 2.90 | 2.79 | 2.71 | 2.64 | 2.59 | 2.54 | 2.40 | 2.33 | 2.20 |
| 16 | 4.49 | 3.63 | 3.24 | 3.01 | 2.85 | 2.74 | 2.66 | 2.59 | 2.54 | 2.49 | 2.35 | 2.28 | 2.15 |
| 17 | 4.45 | 3.59 | 3.20 | 2.96 | 2.81 | 2.70 | 2.61 | 2.55 | 2.49 | 2.45 | 2.31 | 2.23 | 2.10 |
| 18 | 4.41 | 3.55 | 3.16 | 2.93 | 2.77 | 2.66 | 2.58 | 2.51 | 2.46 | 2.41 | 2.27 | 2.19 | 2.06 |
| 19 | 4.38 | 3.52 | 3.13 | 2.90 | 2.74 | 2.63 | 2.54 | 2.48 | 2.42 | 2.38 | 2.23 | 2.16 | 2.03 |
| 20 | 4.35 | 3.49 | 3.10 | 2.87 | 2.71 | 2.60 | 2.51 | 2.45 | 2.39 | 2.35 | 2.20 | 2.12 | 1.99 |
| 21 | 4.32 | 3.47 | 3.07 | 2.84 | 2.68 | 2.57 | 2.49 | 2.42 | 2.37 | 2.32 | 2.18 | 2.10 | 1.96 |
| 22 | 4.30 | 3.44 | 3.05 | 2.82 | 2.66 | 2.55 | 2.46 | 2.40 | 2.34 | 2.30 | 2.15 | 2.07 | 1.94 |
| 23 | 4.28 | 3.42 | 3.03 | 2.80 | 2.64 | 2.53 | 2.44 | 2.37 | 2.32 | 2.27 | 2.13 | 2.05 | 1.91 |
| 24 | 4.26 | 3.40 | 3.01 | 2.78 | 2.62 | 2.51 | 2.42 | 2.36 | 2.30 | 2.25 | 2.11 | 2.03 | 1.89 |
| 25 | 4.24 | 3.39 | 2.99 | 2.76 | 2.60 | 2.49 | 2.40 | 2.34 | 2.28 | 2.24 | 2.09 | 2.01 | 1.87 |
| 26 | 4.23 | 3.37 | 2.98 | 2.74 | 2.59 | 2.47 | 2.39 | 2.32 | 2.27 | 2.22 | 2.07 | 1.99 | 1.85 |
| 27 | 4.21 | 3.35 | 2.96 | 2.73 | 2.57 | 2.46 | 2.37 | 2.31 | 2.25 | 2.20 | 2.06 | 1.97 | 1.84 |
| 28 | 4.20 | 3.34 | 2.95 | 2.71 | 2.56 | 2.45 | 2.36 | 2.29 | 2.24 | 2.19 | 2.04 | 1.96 | 1.82 |
| 29 | 4.18 | 3.33 | 2.93 | 2.70 | 2.55 | 2.43 | 2.35 | 2.28 | 2.22 | 2.18 | 2.03 | 1.94 | 1.81 |
| 30 | 4.17 | 3.32 | 2.92 | 2.69 | 2.53 | 2.42 | 2.33 | 2.27 | 2.21 | 2.16 | 2.01 | 1.93 | 1.79 |
| 40 | 4.08 | 3.23 | 2.84 | 2.61 | 2.45 | 2.34 | 2.25 | 2.18 | 2.12 | 2.08 | 1.92 | 1.84 | 1.69 |

```
370 IF ABS(T)>1E-15 THEN 330
380 RETURN
390 INPUT V
400 IF V<1 THEN 430
410 IF INT(V)<>V THEN 430
420 RETURN
430 PRINT "INVALID DATA"
440 STOP
```

**Table 11.5** Percentage points for the $t$ distribution

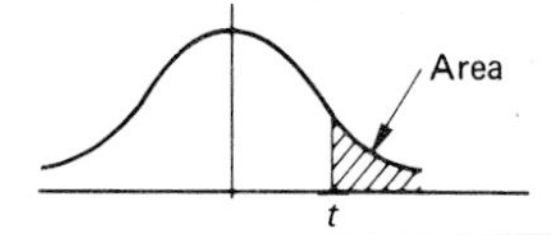


| Degrees of freedom, $\nu$ | Area to the right | | | |
|---|---|---|---|---|
| | 0.05 | 0.025 | 0.01 | 0.005 |
| 1 | 6.314 | 12.706 | 31.821 | 63.657 |
| 2 | 2.920 | 4.303 | 6.965 | 9.925 |
| 3 | 2.353 | 3.182 | 4.541 | 5.841 |
| 4 | 2.132 | 2.776 | 3.747 | 4.604 |
| 5 | 2.015 | 2.571 | 3.365 | 4.032 |
| 6 | 1.943 | 2.447 | 3.143 | 3.707 |
| 7 | 1.895 | 2.365 | 2.998 | 3.499 |
| 8 | 1.860 | 2.306 | 2.896 | 3.355 |
| 9 | 1.833 | 2.262 | 2.821 | 3.250 |
| 10 | 1.812 | 2.228 | 2.764 | 3.169 |
| 11 | 1.796 | 2.201 | 2.718 | 3.106 |
| 12 | 1.782 | 2.179 | 2.681 | 3.055 |
| 13 | 1.771 | 2.160 | 2.650 | 3.012 |
| 14 | 1.761 | 2.145 | 2.624 | 2.977 |
| 15 | 1.753 | 2.131 | 2.602 | 2.947 |
| 16 | 1.746 | 2.120 | 2.583 | 2.921 |
| 17 | 1.740 | 2.110 | 2.567 | 2.898 |
| 18 | 1.734 | 2.101 | 2.552 | 2.878 |
| 19 | 1.729 | 2.093 | 2.539 | 2.861 |
| 20 | 1.725 | 2.086 | 1.528 | 2.845 |
| 21 | 1.721 | 2.080 | 2.518 | 2.831 |
| 22 | 1.717 | 2.074 | 2.508 | 2.819 |
| 23 | 1.714 | 2.069 | 2.500 | 2.807 |
| 24 | 1.711 | 2.064 | 2.492 | 2.797 |
| 25 | 1.708 | 2.060 | 2.485 | 2.787 |
| 26 | 1.706 | 2.056 | 2.479 | 2.779 |
| 27 | 1.703 | 2.052 | 2.473 | 2.771 |
| 28 | 1.701 | 2.048 | 2.467 | 2.763 |
| 29 | 1.699 | 2.045 | 2.462 | 2.756 |
| 30 | 1.697 | 2.042 | 2.457 | 2.750 |
| 40 | 1.684 | 2.021 | 2.423 | 2.704 |
| 60 | 1.671 | 2.000 | 2.390 | 2.660 |

For larger values of $\nu$ treat $t$ as a $z$ score and use the standard normal table..

*The t distribution*
A $t$ statistic is simply the square root of an $F$ ratio with 1 degree of freedom in its numerator. There are several instances of this throughout the book, as in the Heater Example of Chapter 4. It seemed that whenever we had the choice of performing either an $F$ test or a $t$ test they gave identical results because of this close relationship. Having programmed a computer to calculate probabilities and percentage points for the $F$ distribution, therefore, we can use the same programs to generate them for the $t$ distribution. Hence the exact probability in the right-hand tail of the $t$ distribution can be found by entering $t^2$ into the $F$ probability program with 1 degree of freedom in the numerator and the number of degrees of freedom for $t$ in the denominator. The percentage points are obtained by taking the square root of the corresponding $F$ value. Table 11.5 shows some of these which were derived from $F$ values more accurate than those given in the above program. To achieve this accuracy, the error bounds in lines 27∅ and 37∅ were decreased. Naturally the program takes longer to run in that case.

*The Wilcoxon matched-pair signed-ranks statistic*
This was introduced in the Teaching Example of Chapter 4. Differences between matched data values are divided into two groups, those which are positive and those which are negative. The absolute magnitudes of the differences are now ranked in increasing size order. If the two sets of data do come from the same population, then the sum of the ranks for the positive and negative groups will be about the same. We test the smaller of the two sums to see if it is sufficiently small to be significant. It is always, therefore, a one-tail test. The computer program listed here works through all possible ways of dividing the ranks 1, 2, 3, . . . , $N$ into two groups. It keeps a count of how many of those ways produce a smaller sum of ranks which is less than or equal to the given value, $C$. Finally this count is divided by the total number of ways, to obtain the probability of such a sum or a smaller one occurring by chance based on the null hypothesis that the populations are the same.

```
10 DIM I(20)
20 PRINT "WILCOXON SIGNED RANKS PROBABILITIES"
30 PRINT "-----------------------------------"
40 PRINT
50 PRINT "HOW MANY PAIRS OF DATA VALUES ";
60 INPUT N
70 PRINT "SMALLER SUM OF RANKS ";
80 INPUT C
90 F=C+1
100 IF F>(N+1) THEN F=N+1
110 G=N+1
120 FOR N1=2 TO N
130 FOR K=1 TO N1-1
140 I(K)=K
150 NEXT K
160 FOR M=I(N1-1)+1 TO N
170 T=M
180 FOR J=1 TO N1-1
190 T=T+I(J)
200 NEXT J
210 G=G+1
220 IF T<=C THEN F=F+1
```

```
230 NEXT M
240 L=N1-1
250 I(L)=I(L)+1
260 IF I(L)<=N-N1+L THEN 300
270 IF L=1 THEN 340
280 L=L-1
290 GOTO 250
300 FOR P=L+1 TO N1-1
310 I(P)=I(P-1)+1
320 NEXT P
330 GOTO 160
340 NEXT N1
350 W=F/G
360 PRINT "PROBABILITY OF THIS TOTAL OR LESS IS "; W
370 END
```

The program can be used to determine critical values of the smaller of the two rank sums. Various numbers are tried for $C$ until one is found which produces a probability close to the significance level required. Table 11.6 shows such values, together with a normal approximation which is valid for samples of more than 16 pairs.

**Table 11.6** Wilcoxon matched-pair signed-ranks test critical values

| Number of pairs of data values which were ranked | Maximum value smaller sum of ranks can have and still be significant in a one-tail test | |
|---|---|---|
| | Level of significance | |
| | 0.05 | 0.01 |
| 6 | 2 | sample too small |
| 7 | 3 | 0 |
| 8 | 5 | 1 |
| 9 | 8 | 3 |
| 10 | 10 | 4 |
| 11 | 13 | 7 |
| 12 | 17 | 9 |
| 13 | 21 | 12 |
| 14 | 25 | 15 |
| 15 | 30 | 19 |
| 16 | 36 | 24 |

If the number of pairs which were ranked is greater than 16 the $z$ score

$$z = \frac{m(m+1)/4 - T - 0.5}{\sqrt{[m(m+1)(2m+1)/24]}}$$

where $m$ is the number of ranks and $T$ is the smaller sum of ranks, can be used in a one-tail test.

*Spearman's rank correlation coefficient*

In the Dog Show Example of Chapter 4 we applied the formula for the correlation coefficient to two samples of ranks. As ranks are discrete and not normally distributed the significance of the result cannot be tested in the usual way with the $t$ distribution unless the sample is fairly large. The next program systematically enumerates all possible arrangements of the ranks $1, 2, 3, \ldots, N$ and calculates the sum of squared differences each gives rise to when compared with the natural order $1, 2, 3, \ldots, N$. It compares each of these with the given value, $D$. As can be seen from equation (4.20), this is equivalent to examining the rank correlation coefficient for every possible pair of samples of the given size. The reader is warned that because it adopts a direct enumeration approach, the program takes a long time to run. For $N$ equal to 10 it has to count 3 628 800 different rearrangements and calculate a sum of squared differences for each one. This takes over 20 minutes to do by a large computer.

```
10 DIM I(10)
20 PRINT "SPEARMAN'S RANK CORRELATION PROBABILITIES"
30 PRINT "----------------------------------------"
40 PRINT
50 PRINT "NUMBER OF PAIRS N ";
60 INPUT N
70 PRINT "VALUE OF D ";
80 INPUT D
90 F=1
100 FOR L=1 TO N
110 I(L)=L
120 F=F*L
130 NEXT L
140 M=0
150 S=0
160 FOR L=1 TO N
170 S=S+(L-I(L))*(L-I(L))
180 NEXT L
190 IF S<=D THEN M=M+1
200 K=N-1
210 I(N)=1
220 I(K)=I(K)+1
230 IF I(K)<= N THEN 300
240 I(K)=1
250 K=K-1
260 IF K>1 THEN 220
270 I(1)=I(1)+1
280 K=2
290 IF I(1)>N THEN 370
300 FOR J=1 TO K-1
310 IF I(J)=I(K) THEN 220
320 NEXT J
330 K=K+1
340 IF K<=N THEN 300
350 GOTO 150
360 P=M/F
370 PRINT "PROBABILITY OF D VALUE SMALLER THAN THIS "; P
380 END
```

It would be a very dedicated statistician who was prepared to run and rerun the above program on a trial-and-error basis to find percentage points. The strategy adopted to generate Table 11.7 was to calculate the critical value of $D$ from the appropriate $t$ statistic in the large-sample approximation formula. The true value of $D$ is necessarily smaller than this and also must be even, and so working from this initial guess we move downwards two at a time. For instance, the $t$ calculation for $N$ equal to 8 gives $D$ at the 1% level to be 17.74. The next lower possible value to this is 16 but this yields a probability just greater than 1%, so 14 is tried. This turns out to be correct.

**Table 11.7** Spearman's rank correlation coefficient critical values

| Number of pairs of ranks | Critical value of rank correlation coefficient in a one-tail test | |
|---|---|---|
| | Significance level | |
| | 0.05 | 0.01 |
| 6 | 0.829 | 0.943 |
| 7 | 0.714 | 0.893 |
| 8 | 0.643 | 0.833 |
| 9 | 0.600 | 0.783 |
| 10 | 0.564 | 0.746 |

For samples of pairs bigger than 10 in size test

$$t=r\sqrt{\left(\frac{n-2}{1-r^2}\right)}$$

where $r$ is the coefficient and $n$ is the sample size, against the critical one-tail $t$ value with $(n-2)$ degrees of freedom.

## Summary

**1** Statistical analyses can be performed on computers in three different modes. Firstly a purpose-built statistical package of programs with a vocabulary of commands can be used to edit, sort and process the data. Matrix routines are usually incorporated in such packages and so complicated multivariate work can be done. Some packages are very powerful and represent a great help to the practising statistician. Secondly a spreadsheet program, designed originally to facilitate accountancy calculations, allows the user to manipulate the contents of a grid of cells on the screen as if working on a sheet of paper. The quality of the graphs, bar charts and the like produced by these commercial packages is often higher than those from statistics ones. They are thus especially good for preparing and typesetting reports and other presentations. The third way statistics can be implemented on a computer is to run specific programs to do specific tasks, like generating critical values of test statistics.

**2** A lot of multivariate statistics can be expressed in terms of matrix algebra. The normal equations of multiple regression are a system of linear simultaneous equations and so this is hardly surprising. A principal components analysis corresponds to a rotation of coordinate axes to obtain maximally variable quantities. This time eigenvalues and eigenvectors are needed, so again matrices appear quite naturally.

**3** All the probability distributions of parametric and nonparametric test statistics can be calculated by computer programs. For the distributions of classical sampling theory the relevant integrals are expressed as power series which the computer can sum. For nonparametric applications it is often a matter of enumerating all possibilities to assess the significance of the particular sample value measured. Once a program exists to generate probabilities it can be used on a systematic trial-and-error basis to give critical values of the test statistic.

## Further reading

References (58) and (63) of the Bibliography are introductory manuals to specific statistical packages. Most of the books in the Multivariate Techniques section develop the matrix approach to multiple regression. As it is so commonly available as a computer routine many statisticians use it to perform analyses of variance. They introduce dummy variables which are zero or one and denote whether a particular $y$ value belongs to a particular sample or not. By regressing on these dummy variables it can be seen how significantly different the samples are. Reference (37) contains listings of programs in the FORTRAN language to implement many useful clustering algorithms, while (55), (56) and (57) have simple programs for calculating means, variances, test statistics and regressions. For more comprehensive tables of critical values than those given here the reader is referred to (66), while (61) examines many issues covered in this chapter.

## Exercises

**1** If you have access to a statistical package, try to reproduce the analysis of one or more of the worked examples in other chapters. Obtaining results which are known to be correct will reinforce your confidence in using the computer.
**2** If you have access to a spreadsheet package, try to reproduce the analysis of one or more of the worked examples in the other chapters. This is harder than using a statistical package as you will have to program the cells for yourself. Start with something relatively easy like calculating the mean and variance of a single small sample.
**3** Write a program in any computer language you know to select a random sample of 8 numbers from the computer's random number generator. Calculate the mean of the sample at the end of the program. Put your program into a repetitive loop so that 500 samples of 8 numbers are chosen. Verify that their means are approximately normally distributed in accordance with the central limit theorem and the Law of Large Numbers as discussed in Chapters 1 and 3.

# Bibliography


## General statistics

**1** Chatfield, C., *Statistics for Technology*, Chapman and Hall (3rd edn), 1983.
**2** Clarke, G. M., *Statistics and Experimental Design*, Edward Arnold (2nd edn), 1980.
**3** Clarke, G. M. and Cooke, D., *A Basic Course in Statistics*, Edward Arnold (2nd edn), 1983.
**4** Dixon, W. J. and Massey, F. J., *Introduction to Statistical Analysis*, McGraw-Hill (4th edn), 1983.
**5** Ehrenberg, A. S. C., *A Primer in Data Reduction*, Wiley, 1982.
**6** Jenkins, G. W. and Slack, J. L., *Statistics and Probability*, Heinemann, 1985.
**7** Moroney, M. J., *Facts from Figures*, Pelican Original, Penguin Books (2nd edn), 1953.
**8** Mulholland, H. and Jones, C. R., *Fundamentals of Statistics*, Butterworths, reprinted 1982.
**9** Snedecor, G. W. and Cochran, W. G., *Statistical Methods*, The Iowa State University Press (6th edn), 1967.
**10** Spiegel, M. R., *Probability and Statistics*, Schaum Outline Series, McGraw-Hill, 1980.
**11** Wonnacott, R. J. and Wonnacott, T. H., *Introductory Statistics*, Wiley, 1985.

## Statistics for specific disciplines

**12** Armitage, P., *Statistical Methods in Medical Research*, Blackwell Scientific Publications, Wiley, reprinted 1973.
**13** Bland, J. A., *Statistics for Construction Students*, Construction Press, 1985.
**14** Ferguson, G. A., *Statistical Analysis in Psychology and Education*, McGraw-Hill (5th edn), 1981.
**15** Kazmier, L. J., *Statistical Analysis for Business and Economics*, McGraw-Hill (3rd edn), 1978.
**16** Mather, K., *Statistical Analysis in Biology*, Chapman and Hall, 1972.
**17** Miller, J. C. and Miller, J. N., *Statistics for Analytical Chemistry*, Wiley, 1984.
**18** Parker, R. E., *Introductory Statistics for Biologists*, Edward Arnold, 1973.
**19** Steel, R. G. D. and Torrie, J. H., *Principles and Procedures of Statistics, A biometrical approach*, McGraw-Hill (2nd edn), 1981.

## Theoretical statistics

**20** Box, G. E. P. and Tiao, G. C., *Bayesian Inference in Statistical Analysis*, Addison Wesley, 1973.
**21** Brownlee, K. A., *Statistical Theory and Methodology in Science and Engineering*, Wiley (2nd edn), 1965.

**22** Hoel, P. G., *Introduction to Mathematical Statistics*, Wiley, 1984.
**23** Hogg, R. and Craig, A., *Introduction to Mathematical Statistics*, Macmillan, 1970.
**24** Kendall, M. and Stuart, A., *The Advanced Theory of Statistics*, Griffin, 3 volumes, 1987, 1979 and 1983.
**25** Lindgren, B. W., *Statistical Theory*, Macmillan (2nd edn), 1968.
**26** Lipschuts, S., *Probability*, Schaum Outline Series, McGraw-Hill, 1974.
**27** Maritz, J., *Empirical Bayes Method*, Methuen, 1970.
**28** Mood, A. M., Graybill, F. A. and Boes, D. C., *Introduction to the Theory of Statistics*, Wiley (3rd edn), 1974.

## Nonparametric methods

**29** Conover, W. J., *Practical Non-parametric Statistics*, Wiley (2nd edn), 1980.
**30** Hollander, M. and Wolfe, D. A., *Nonparametric Statistical Methods*, Wiley, 1973.
**31** Leach, C., *Introduction to Statistics*, Wiley, 1979.
**32** Maxwell, A. E., *Analysing Qualitative Data*, Chapman and Hall, 1961.
**33** Siegal, S., *Nonparametric Statistics for the Behavioural Sciences*, McGraw-Hill, 1956.

## Multivariate techniques

**34** Chatfield, C. and Collins, A. J., *Introduction to Multivariate Analysis*, Chapman and Hall, 1980.
**35** Draper, N. R. and Smith, J., *Applied Regression Analysis*, Wiley, 1966.
**36** Gilchrist, W., *Statistical Modelling*, Wiley, 1984.
**37** Hartigan, J. A., *Clustering Algorithms*, Wiley, 1975.
**38** Kendall, M. G., *Multivariate Analysis*, Griffin (2nd edn), 1980.
**39** Maxwell, A. E., *Multivariate Analysis in Behavioural Research*, Chapman and Hall, 1977.
**40** Morrison, D. F., *Multivariate Statistical Methods*, McGraw-Hill, 1967.
**41** Sneath, P. H. A. and Sokal, R. R., *Principles of Numerical Taxonomy*, Freeman, 1963.

## Experimental design

**42** Cochran, W. G. and Cox, G. M., *Experimental Designs*, Wiley (2nd edn), 1957.
**43** Cox, D. R., *Planning of Experiments*, Wiley, 1958.
**44** Johnson, N. L. and Leone, F. C., *Statistics and Experimental Design in Engineering and the Physical Sciences*, Wiley (2nd edn), 1977.
**45** Scheffé, H., *The Analysis of Variance*, Wiley, 1959.

## Questionnaire design and sample surveys

**46** Churchill, G. A., *Marketing Research*, Dryden Press (2nd edn), 1979.
**47** Crouch, S., *Marketing Research for Managers*, Heinemann, 1984.
**48** Gardner, G., *Social Surveys for Social Planners*, Open University Press, 1978.
**49** Government Statistical Service, *Government Statistics – A Brief Guide to Sources*, Her Majesty's Stationery Office, annually.
**50** Harvey, J. M., *Sources of Statistics*, Clive Bingley, 1971.
**51** Hoinville, G. and Jowell, R., *Survey Research Practice*, Heinemann, 1978.

**52** Moser, C. and Kalton, G., *Survey Methods in Social Investigation*, Heinemann (2nd edn), 1974.

## Miscellaneous topics

**53** Blower, J. G., Cook, L. M. and Bishop, J. A., *Estimating the Size of Animal Populations*, Allen and Unwin, 1980.
**54** Caplen, R. N., *A Practical Approach to Quality Control*, Business Books (4th edn), 1982.
**55** Cohen, S. S., *Operational Research*, Edward Arnold, 1985.
**56** Cooke, D., Craven, A. H. and Clarke, G. M., *Basic Statistical Computing*, Edward Arnold, 1982.
**57** Cooke, D., Craven, A. H. and Clarke, G. M., *Statistical Computing in Pascal*, Edward Arnold, 1985.
**58** Higginbotham, P. G., *Microtab - An All-purpose Statistical Package*, Edward Arnold, 1985.
**59** Kelejian, H. H. and Oates, W. E., *Introduction to Econometrics*, Harper and Row, 1981.
**60** Kendall, M. G., *Time Series*, Griffin (2nd edn), 1976.
**61** Maindonald, J. H., *Statistical Computation*, Wiley, 1984.
**62** Murdoch, J., *Control Charts*, Macmillan, 1979.
**63** Ryan, T. A., Joiner, B. L. and Ryan, B. F., *Minitab - Student Handbook*, Duxbury Press (2nd edn), 1985.
**64** Wetherill, G. B., *Sampling Inspection and Quality Control*, Chapman and Hall (2nd edn), 1977.
**65** Wheelwright, S. C. and Makridakis, S. G., *Forecasting Methods for Management*, Wiley (3rd edn), 1980.
**66** White, J., Yeats, A. and Skipworth, G., *Tables for Statisticians*, Stanley Thornes (3rd edn), 1979.

# Answers to Exercises

## Chapter 1

**3** (i) metric, discrete, (ii) nominal, (iii) metric, discrete, (iv) metric, continuous, (v) ordinal, (vi) metric, continuous, (vii) ordinal, (viii) nominal. **4** His 'bomb' is not really a bomb. **5** 4/7. **6** (i) 0.16, (ii) 0.32, (iii) 0.64.

## Chapter 2

**2** (ii) 15.83 years, 14.5 years, 17 years, 42.97 $\text{years}^2$, 6.55 years. **3** (i) Frequencies are 6, 36, 33, 4 and 2, (iii) 65.728 dB, 17.600 $\text{dB}^2$, (iv) 64.91 dB, 21.319 $\text{dB}^2$, (v) Answers fairly close and so the classification is acceptable. **4** (ii) Approximately 460 seconds, 520 seconds, 360 seconds, 80 seconds, (iii) approximately 0.48, (iv) approximately 0.39. **5** (ii) £45 325, £1414 (iii) 103, 109, 107, 109, 108, 108, 109, (iv) There seems to be no relationship between advertising and sales.

## Chapter 3

**1** (i) 0.68, (ii) 0.45, (iii) 74 mm to 136 mm, (iv) 94.5 mm to 115.5 mm. **2** (i) 3.19 man-hours to 6.77 man-hours, (ii) $t=-1.291$, not significant at 5% in two-tail test, no evidence that population mean is different from 6 man-hours, (iii) $t=1.8730$, significant at 5% in one-tail test, population mean is bigger than 3.5 man-hours, (iv) 2.95 $\text{man-hours}^2$ to 20.81 $\text{man-hours}^2$. **3** $\chi^2=9.743$, not significant at 5%, no evidence that the die is unfair. **4** (i) 116, (ii) 0.3946, (iii) 0.3666, 0.4778, 0.1557, (iv) 53.89, 70.24, 22.89, (v) $\chi^2=0.066$ using Yates' correction, not significant at 5% and so the Hardy-Weinberg Law is valid. **5** Taking '4 or more' to be equal to 5, the mean is 1.6992 beds per week. Expected frequencies are 24.32, 41.32, 35.10, 19.88 and 12.38 giving $\chi^2=1.334$. This is not significant at 5% and so the Poisson distribution does describe the data. **6** (i) 0.337 to 0.389, (ii) $z=4.122$, significant at 5% in one-tail test, population proportion is bigger than 0.3. **7** (i) 13 to 29 correct guesses, (ii) $z=1.816$, significant at 5% in one-tail test; she is not guessing, (iii) Examine observed and expected frequencies for correct and incorrect guesses. **8** $P$(5 or less below median) = 0.212, not significant at 5%, median could be £280.

## Chapter 4

**1** (i) $\chi^2=12.0306$, not significant at 5%, no association, (ii) Taking expected totals in the ratio 1:2:1 gives $\chi^2=100.13$ which is highly significant; there is an association between the type of crime committed per head of the population and the country, (iii) $\chi^2=19.12$, highly significant; there is a significant difference between the numbers of traffic offences per head of population in the three countries. **2** (ii) $y=$

$-0.226x+116.603$, (iii) $-0.6916$, (iv) 82.38% to 98.15%, (v) 90.38% to 97.07%. **3** (i) 22.57 minutes, 2.793 minutes, (ii) 20.96 minutes to 24.18 minutes, (iii) $t=2.105$, not significant at 5% in two-tail test, no evidence that population mean is different from 21 minutes, (iv) $t=2.500$, significant at 5% in one-tail test; afternoon times are significantly longer than morning ones, (v) Rank sum = 1.5, significant at 5%, afternoon times bigger, (vi) $P$(at most 1 reversal of times out of 6 carpenters) = 0.109, not significant at 5%, no difference in job times. (Note that the sign test is not as sensitive as the others.) **4** Correlation coefficient = 0.9836, hence the shape is a catenary. $A$ and $B$ are approximately 0.18 and 0.68 respectively.

## Chapter 5

**1** Case profiles should, roughly speaking, identify different schools of painters. **2** Clusters are ABC, GIJ, D, E, F, H. **3** sales = 51.93 − 1.56 × (advertising in previous year) + 0.25 × (year), $R^2=0.64$. **4** $y=16.6+5.9x_1-4.3x_2$, $R^2=0.72$, $y$ on $x_1$ only gives $y=8.1+5.4x_1$ with $R^2=0.58$, $y$ on $x_2$ only gives $y=35.3-2.8x_2$ with $R_1^2=0.06$, $x_2$ can be omitted.

## Chapter 6

**1** (i) $\chi^2=4.50$ using Yates' correction, there is an association and so the pesticides differ, (ii) $z=2.1223$ for difference in proportions dying, significant at 5% in one-tail test so Pesticide A is better than Pesticide B. **2** (i) $T=118$, $z=7.96$, highly significant in one-tail test, experienced rats are better, (ii) $t=5.818$, significant at 5% in one-tail test, same conclusion as in (i). **3** $F$ ratio between residual mean squares of individual regressions is 1.82, hence background variances are the same. $F$ ratio between individual slopes and pooled slope (0.31) is 0.25, hence common slope is acceptable. $F$ ratio between different intercepts and common intercept is 6.10 which is significant at 5%, hence intercepts are significantly different.

## Chapter 7

**1** $F=33.8512$ which is 5.818 squared. **2** (i) $F=4.488$, treatments are significantly different, (ii) $t=1.213$, Eatwell and Jerkaround are not significantly different, (iii) $t=2.468$, Eatwell is significantly different from the mean of Jerkaround and Sleepalot, (iv) Ratio of maximum $L$ to $s_L$ is 2.98 while the critical value is 2.71, hence Sleepalot and Eatwell are significantly different even though they are self-selecting in giving the maximum difference. **3** (i) Types of resistor and temperatures give significant effects, (ii) $t=2.96$, significant at 5%, two groups are different, (iii) Linear contrast is $-52$, quadratic is $-22$ with $F$ ratios 31.87 and 4.08 respectively. Hence temperature effect is linear with non-significant quadratic and higher order effects. **4** $F$ ratios for states, months and interaction are 4.69, 1.91 and 0.98 respectively. Hence state is significant but not month and there is no significant interaction.

## Chapter 8

**1** $F=2.9826$ which is 1.727 squared. **2** (i) $F$ for subjects is 10.50 which is significant, hence subjects are significantly different. (ii) $t=0.67$, not significant in two-tail test, English and Geography marks are not significantly different, (iii) $F=5.76$, subjects still significantly different. **3** $F$ ratios for fuel, injection rate and engine effects are 4.89, 5.82 and 4.73 respectively, all factors are significant. **4** Family size is significant with $F$ ratio 8.22 but no other effects are. **5** Trawlers are significantly different with $F$ ratio 4.62 but type of music is not significant.

## Chapter 9

**4** (i)$\chi^2 = 13.00$, brands do differ, (ii) $\chi^2 = 1.805$, using Yates' correction, data is consistent with claim, (iii) $\chi^2 = 10.812$, there is an association. **5** (i) $\chi^2 = 8.464$, respondents are not choosing their first choice randomly, (ii) Expected frequencies are 47.5 and 58, (iii) $\chi^2 = 39.55$, second choices are not being made at random.

## Chapter 10

**1** (i) Mean and standard deviation are approximately 17.55 mg per litre and 1.97 mg per litre respectively, (ii) about 37%. **2** (i) $m$ and $\lambda$ are approximately 2.18 and 0.000 061 respectively, (ii) about 17%. **3** 95% limits are 11.62 mm to 12.38 mm for the mean and 0.50 mm to 2.05 mm for the range, 99% limits are 11.51 mm to 12.49 mm for the mean and 0.35 mm to 2.36 mm for the range. **4** $P$(0 or 1 defective) is 0.962 which is greater than 95% and so only 1 defective should be tolerated. **5** (i)——, (ii) Producer's risk is 0.033% and the consumer's risk is 44.5%, (iii) Producer's risk is 0.198% and the consumer's risk is 48.3%, hence both parties are worse off with the new scheme.

# Index